Agricultural Policies for Poverty Reduction

Edited by Jonathan Brooks

OECD

Please cite this publication as:
OECD (2012), *Agricultural Policies for Poverty Reduction*, OECD Publishing.
http://dx.doi.org/10.1787/9789264112902-en

ISBN 978-92-64-16863-3 (print)
ISBN 978-92-64-11290-2 (PDF)

CABI ISBN 978-17-80-64105-8 (hardback)

Photo credits: Cover illustration © Shutterstock/Worakit Sirijinda

Corrigenda to OECD publications may be found on line at: *www.oecd.org/publishing/corrigenda*.

Foreword

The chapters in this volume are concerned with the role of agricultural policies in developing countries, with a focus on their potential for raising the incomes of agricultural and rural households and thereby reducing poverty.

More than two-thirds of the world's poor live in rural areas, and higher rural incomes are a pre-requisite for sustained poverty reduction and the attainment of global food security.

In recent years, there has been renewed recognition of the role that agricultural development has to play in poverty reduction, which has led to commitments to reverse a 30-year decline in development funding for the sector. As public investments are scaled up, there is a pressing need to ensure that appropriate policies are put in place.

Progress in reducing rural poverty has been uneven. In global terms, the proportion and total number of people living in absolute poverty has declined, with many countries – most notably China – showing rapid improvements. But in parts of Africa and South Asia, progress has been fitful or non-existent.

The policies required to boost rural incomes comprise those that can raise farm household incomes directly, complemented by those that can smooth the transition to a more diversified rural economy. The precise mix will depend on many factors, including a country's stage of economic development and its resource endowments.

The majority of the world's poor now live in lower middle-income countries, where the process of growth, development and poverty reduction is underway. In those countries a major challenge lies in connecting poor rural households to that development mechanism as efficiently as possible. That means recognising that – as OECD countries have all experienced – the majority of future generations will have better economic prospects outside agriculture than within it. In many poorer countries, smallholder agriculture remains the backbone of the rural economy and improvements in farm productivity are an essential stepping stone to broader economic development.

Agricultural policies are just one element of the overall mix of policies, investments and institutional reforms that are often needed. In the case of OECD countries, our analysis has made clear that market interventions, such as price controls and subsidies for fertiliser and other inputs, tend to be inefficient at raising farmers' incomes, inequitably distributed, and impose negative spill-overs on other countries. OECD analysis has concluded that policies should instead provide targeted income assistance to poorer households via social programmes and invest in public goods, such as research and development and rural infrastructure, in order to improve the competitiveness of the sector. It is also essential to facilitate the transition to more remunerative activities when farm operations are not inherently viable.

In poorer countries, this is also a preferred strategy, but there may be problems implementing it. Under-developed institutions and endemic market failures have therefore led to "second best" options being explored, including the use of price stabilisation schemes and subsidies for fertiliser and other inputs. This volume considers the specific circumstances under which the use of such instruments might be warranted.

The analysis comes at an important time. High agricultural prices have raised the living costs of many poor households, but also provide an opportunity for farmers in developing countries. For those opportunities to be exploited, a variety of constraints – human, natural resources and environmental – will need to be overcome. It is also important that the benefits of growth are shared widely. Trade has an important role to play, with further market opening, accompanied by effective transition and adjustment assistance, offering widespread benefits to both developed and developing countries. These are complex challenges, but the analysis in this volume provides important insights into the specific role of agricultural policy and how that role may vary for countries at differing stages of development.

Ken Ash
Director, Trade and Agriculture Directorate

Acknowledgements

This book is the outcome of an OECD project on Agricultural policy choices in developing countries. *The material forms the basis for a shorter policy synthesis,* Agricultural policies for poverty reduction: A synthesis, *which was declassified by the OECD's Working Party on Agricultural Policies and Markets in June 2011. Specific acknowledgements are made at the end of each country chapter. In addition, the editor, Jonathan Brooks, would like to acknowledge the contributions of Florence Bossard and Clarisse Legendre, who provided statistical support, and Anita Lari and Michèle Patterson, who prepared the material for publication.*

Contributors

Philip C. Abbott *is a Professor of Agricultural Economics at Purdue University. His current research focuses on cotton and cocoa in West Africa, on trade and development in Viet Nam, on volatile food prices in international markets, and on trade and stabilisation policy responses by developing countries.*

Jonathan Brooks *is a Senior Economist in the OECD Trade and Agriculture Directorate and has worked mainly in the areas of agricultural policy, trade and development. His current research focuses on food security and the role of agricultural policy in poverty reduction.*

Katia Covarrubias *is an economist in the Agricultural Development Economics Division of the UN Food and Agriculture Organisation. She has contributed to the development and expansion of the* Rural Income Generating Activities Database. *Her current research interests include income generation strategies, food security, and the economic impacts of cash transfer programs.*

Mateusz Filipski *is a Post-Doctoral Researcher in the department of Agricultural and Resource Economics of the University of California, Davis. His current research focuses on economy-wide modelling methods applied to policy analysis in developing countries.*

Erik Jonasson *is a Post-Doctoral Researcher at the Department of Economics, Lund University, Sweden. His research focuses on labour market issues in developing countries, with an emphasis on rural non-agricultural employment.*

J. Edward Taylor *is Professor of Agricultural and Resource Economics and Director of the Center on Rural Economies of the Americas and Pacific Rim (REAP) at the University of California, Davis. He has written extensively on the economy-wide impacts of agricultural and development policies.*

Steve Wiggins *is a Research Fellow at the Overseas Development Institute. He has been studying and working on agricultural and rural development in Africa and Latin America since 1972. His main interests are in rural livelihoods, poverty, food security and nutrition.*

Table of Contents

Tables

Figures

Executive Summary

This study addresses the role of agricultural policies in raising incomes in developing countries. Higher incomes are essential for sustained progress on the first Millennium Development Goal (MDG1), which calls for the eradication of extreme poverty and hunger, and includes a specific target of reducing by 50% between 1990 and 2015 the proportion of people living on less than a dollar a day. The aim is to identify ways in which the appropriate set of policies may vary according to a country's stage of development.

In addressing this question, it is helpful to make a distinction between policies that are likely to be effective in the short- to medium-term, a period which now encompasses the timeframe of MDG1, and those that are likely to produce fundamentally stronger incomes in the long term. In the short to medium term economic structures are fixed, while in the long term they can evolve.

In the short to medium term, there is a need to raise the basic incomes of the poor and to strengthen systems of social protection. Since over two-thirds of the world's dollar-a-day poor live in rural areas, where farming is a core economic activity, this implies policies and investments that raise economic returns within agriculture.

In the long run, there is a need to anticipate the structural changes in agriculture that accompany successful economic development. These include *i)* a declining share of agriculture in GDP as the economy develops and diversifies; *ii)* a release of labour from the sector driven by a combination of the "push" of labour-saving technical change in agriculture and the "pull" of growing labour demand in non-agricultural sectors; and *iii)* rising agricultural output.

This means offering multiple development pathways for farm households: improving competitiveness (*i.e.* productivity) within the agricultural sector; diversifying income sources among household members; and, for some, leaving the sector for better paid jobs.

Within agriculture, there are rising opportunities for smallholders, who typically underpin developing country agriculture, to establish commercially viable operations. Yet exploitation of those opportunities will itself imply some less productive smallholders moving on to other, ultimately more remunerative, activities. Even in poor agriculture-dependent economies there is thus a need to strengthen opportunities outside as well as within farming.

Many of the policies required to improve farmers' opportunities are non-agricultural. Improvements in education and primary healthcare are key to prospects within and outside the sector. Equally important is the overall investment climate, which depends on factors such as peace and political stability, sound macroeconomic management, developed institutions, property rights and governance. In agriculture-dependent economies, there is a strong case for increasing the share of public spending allocated to sectoral public goods, such as rural roads and agricultural research, and to technology transfer, farm extension and advisory services.

This long-term approach is consistent with the overall policy recommendation for OECD countries, which suggests that governments can promote long-term agricultural development by reducing market failures, ideally by tackling them at source, and by providing public goods. It leaves an important role for targeted social policies in helping farmers who cannot adjust within the current generation, and for addressing immediate concerns about the level and distribution of income.

In low-income countries, however, it has been suggested that – because of weak institutions and endemic market failures – market interventions might also be warranted. Price support, price stabilisation, and input subsidies have been proposed as ways of addressing short-term objectives with respect to incomes, poverty and food security, and of promoting long-term economic development.

In the short term, price policies provide an easy lever for government, but are inefficient at addressing income concerns. Price support for food products is a blunt instrument because, among the poor, there are net sellers and net buyers of food – in many poor countries, the majority of farm households are net buyers. Price stabilisation (as opposed to price support) can limit the impact of adverse shocks on producers and consumers, but often proves to be fiscally unsustainable. A preferable option for the poor – both producers and consumers – is targeted social programmes, including cash transfers, although these may be difficult to implement in the poorest economies. At the same time, agricultural investments can improve farmers' resilience to risk.

Over the long term, market interventions treat the symptoms of market failure and under-development rather than the causes. Price stabilisation can provide a more stable investment climate, but thwarts the development of private risk management and can export instability onto world markets. Input subsidies can redress failings such as the under-development of infrastructure, missing markets for credit and inputs, and a lack of knowledge of the benefits of using improved seed and fertiliser, but can impede the development of private markets. In both cases, the benefits and costs of intervention need to be judged relative to the benefits and costs of tackling the underlying problems directly.

Finally, there are dangers in using market interventions to address multiple economic and social objectives. Such programmes can become an easy target for interest groups, outliving their original justification and becoming a budgetary millstone. An important priority is that expenditures on market interventions should not crowd out essential investments in support of long-term agricultural development.

Agricultural policies for raising rural incomes: An introduction

by

Jonathan Brooks

The importance of raising agricultural and rural incomes in developing countries is high on the political agenda, for both structural reasons and as a result of recent developments in world food markets.

The fundamental reason is the persistence of poverty and hunger. While the incidence of dollar-a-day poverty has declined substantially in recent decades, population growth has implied a much more modest reduction in the total number of poor. Progress has also varied by region. In South Asia, the total number of poor has been more or less constant, while in Africa the total is actually increasing. Globally, three out of every four poor people live in rural areas – corresponding to 2.1 billion people living on less than USD 2 a day and 880 million living on less than USD 1 a day (World Bank, 2007). Raising their incomes is essential for reduced poverty and improved food security.

The renewed focus on agricultural and rural development has also been given impetus by the recent food price crisis. The FAO estimates that the food price spike in 2007-08 caused the number of undernourished people in the world to reach 1.02 billion in 2009, compared with an average of 850 million in 2005-07. The estimated total for 2010 is 925 million, considerably higher than the pre-crisis level, which had shown little change over a 30-year period. While the biggest increases in food bills were incurred by urban consumers, the most pervasive and devastating impacts on food security were among the rural poor, including farm households, the majority of whom are net buyers of food (Zezza *et al.*, 2008).

The concern with high food prices contrasts with decades of emphasis on the adverse impacts of low food prices on developing country farmers, motivated by the 1950 Prebisch-Singer hypothesis that the terms of trade between primary products (notably food) and manufactures will tend to decline over time. For decades, the decline in real agricultural prices appeared to confirm this hypothesis, as productivity and supply increases outpaced demand growth. World prices were also suppressed by the effects of farm subsidies in OECD countries. These effects prompted concerns about the real incomes of less competitive farmers in developing countries (particularly smallholders), who were not responsible for generating productivity improvements and were thus caught in a price-cost squeeze, and who furthermore faced unfair competition from farmers in OECD countries.

Is it inconsistent to emphasise the negative impacts of both low and high prices? Not really. The direct effect of price movements on farmers is by nature ambiguous, with net sellers tending to gain and net buyers to lose. Moreover, some farmers may have been hurt by both low and high prices; the former causing them to sell assets, reduce production and retreat into autarky or become net buyers of food; the latter raising the cost of food purchases. The *OECD/FAO Agricultural Outlook* suggests that, while agricultural commodity prices are likely to fall back from recent highs, they are also likely to remain higher in the coming decade than in the previous one (OECD and FAO, 2011). Insofar as stronger prices do not simply reflect higher input costs, the positive long-term outlook offers opportunities for developing country farmers, provided they can respond to these improved incentives.

The factors driving the recent spike in food prices were complex. They included market fundamentals, with drought in Australia and unfavourable harvests in Russia, Ukraine and the United States contrasting with continued demand growth in the developing world. The spike was aggravated by low stocks and by policies adopted in many countries, including export restrictions and the hoarding of commodities. Further contributing factors were biofuel policies, which diverted agricultural supplies from food to energy uses, and the increased link of food to energy prices, which reflected both the use of agricultural products in biofuels and rising direct energy costs. There is also a lively debate over the extent to which short-term price movements may have been amplified by speculators on futures markets (OECD, FAO and others, 2011).

Many of the factors contributing to high food prices are also likely to be associated with increased price volatility. If supply conditions are already tight, then a supply shock is likely to result in a bigger upward price movement. The increased link of commodity prices to fuel prices is also likely to exacerbate price volatility, as is growth of production in regions where yields are less stable. If governments act to stabilise domestic markets, then that is likely to export instability onto international markets. Speculation should in principle stabilise markets by allowing for price discovery, but some have argued that it has had the opposite effect. As prices fell and then rose again in 2010, price volatility and its implications for food security became an overriding concern, prompting G20 leaders to request a group of international organisations to examine these linkages and propose policy options for managing risk and protecting the most vulnerable. As a result, agriculture ministers agreed on an action plan which includes improving agricultural production and productivity; increasing market information and transparency; strengthening international policy co-ordination; improving risk management tools for governments, firms and farmers; and improving the functioning of agricultural commodities' derivative markets. The basic aim is to build more resilient farm structures and rural economies.

The emphasis on improving the productivity of agriculture underscores that the basic problem of low agricultural incomes goes far beyond the level or volatility of prices. Even in the poorest countries, it is rare for agriculture to account for more than a third of GDP. Yet agriculture often accounts for two-thirds or more of employment in low income countries. The wide gap between agriculture's share of GDP and the sector's share of employment, apparent in nearly all developing countries, reflects to a considerable extent the relatively low productivity of labour in agriculture and consequently low remuneration. This basic problem predates the food crisis.

The most effective way to narrow the gap between agricultural and non-agricultural incomes is a matter of some controversy. One line of argument runs that, in the poorer countries of Africa and South Asia, development strategies should focus on improving agricultural productivity directly, with policies and public investments focusing on agriculture in general and on smallholder development in particular (Hazell *et al.*, 2007; Morris *et al.*, 2009). The fundamental reasoning behind this argument is that smallholder farming underpins rural economies and forms the basis for the livelihoods of the world's poor, and the development process has to start by tapping the potential of existing structures. The counter argument is that the necessary improvement in labour productivity will require a vast reduction in the proportion of the population engaged in agriculture, and that policy should focus on promoting competitive agriculture and facilitating the exit of non-competitive farmers from the sector (Collier and Dercon, 2009).

There is merit in both arguments. The analysis in this volume seeks to promote balance in this discussion, making a distinction between policies that are likely to be effective in raising incomes in the short- to medium-term, a timeframe in which economic structures are essentially fixed, and those that have the potential to generate fundamentally stronger incomes in the long term, by facilitating the transition to more efficient economic structures. Chapter 1 provides a framework for strengthening rural incomes, based on the premise that it is essential for policy makers to acknowledge the structural changes to agriculture that accompany successful economic development. These include *i)* a declining share of agriculture in GDP as the economy develops and diversifies; *ii)* a release of labour from the sector driven by a combination of the "push" of labour-saving technical change in agriculture and the "pull" of growing labour demand in non-agricultural sectors; and *iii)* rising agricultural output. The third factor emphasises that agriculture typically declines in relative but not absolute terms.

A key point is that a strong market outlook should imply rising opportunities for smallholder farmers, and it is important to raise productivity for those gains to be realised. At the same time, however, the exploitation of those opportunities by some smallholders will inevitably lead to other smallholders leaving the sector. The main reasons for this are twofold. First, very small operations, often of half a hectare or less, are typically too small to benefit from economies of scale in areas such as procuring inputs, obtaining information on markets and technical issues, in meeting standards and certifying production, and in transacting with buyers from processors and supermarkets. Second, beyond questions of efficiency, the smallest farms are in most cases incapable of generating a level of income which would keep the household much above the poverty line. As a consequence, policy makers need to adopt a balanced strategy that enables households to improve their competitiveness (*i.e.* productivity) within agriculture, yet also makes it possible for them to diversify their income sources and be in a position to take better paid jobs outside the sector. The broad requirement is for agriculture to be situated within a comprehensive rural development strategy, so that the demand for labour from other sectors matches the release of labour induced by productivity improvements – *i.e.* labour is "pulled" rather than "pushed" out.

The objective of raising rural incomes needs to be seen in the wider context of global and national challenges faced by policy makers. At the global level, these include feeding a rising world population which is expected to exceed nine billion by 2050, and doing so in a manner that is environmentally sustainable and compatible with agreed targets for greenhouse gas emissions. At the national level, concerns with respect to incomes are part of the wider agenda for poverty reduction and food security, and sit alongside other objectives, such as the sustainable use of resources and the promotion of gender equality. The focus of the remaining chapters is on the specific role of agricultural policy instruments in addressing income-related concerns, recognising that other objectives may also be important.

Developing countries differ widely in terms of their natural resource endowments, the types of agriculture systems that are in place, farm sizes and land tenure laws, basic levels of human development in areas such as health and education, and the development of government and administrative institutions. A specific concern is that markets are less developed than in higher income countries. For example, transaction costs may be higher in output markets, meaning that farmers are less engaged with markets; some markets (*e.g.* for credit and insurance) may be missing altogether and market failures (*e.g.* arising from insecure property rights, incomplete information, or market power) may be more endemic.

The appropriate mix of policies is likely to vary according to a country's structural circumstances and its stage of economic development. The opportunities and constraints facing a subsistence farmer in Bangladesh may look superficially similar to those facing a peasant farmer in neighbouring India. But the balance of opportunities within and outside agriculture, and the constraints to overcoming them, may be fundamentally different.

OECD countries agree on some basic principles with respect to the pursuit of income-related objectives in member countries (OECD, 2002): countries should use social policies to protect incomes in the short term (and provide support for farmers who are unable to adjust), while correcting market failures and investing in public goods in order to strengthen agricultural incomes more fundamentally over the medium to long term. This approach contrasts with using market distorting interventions, such as price supports and input subsidies, which are found to perform poorly in terms of raising the incomes of farm households (OECD, 2001) and often have perverse distributional effects, paying more to larger and richer farmers than to smaller and poorer ones, and taking money away from consumers and taxpayers to boost the incomes for households whose incomes are already above average (OECD, 2003).

In the case of poorer countries, however, it has been argued that this advice might need to be qualified. In the first place, developed country systems of social protection may not be in place and – pending their development – market interventions may be the only practical way of responding to events such as the recent spike in world food prices. Second, market failures are likely to be more endemic and it may be difficult to tackle them directly. For example, farmers may have low incomes partly because they have no access to credit. Input subsidies have thus been suggested as a practical solution to the otherwise difficult problem of developing input markets and providing financial services to small farmers. Similarly, price stabilisation has been proposed as a relatively simple way of mitigating the impacts of price shocks on poor households, as opposed to market-based forms of risk management or the provision of income safety nets. These specific arguments are taken up in later chapters.

Chapter 2 considers the impact of price changes on the real incomes of rural households. The approach uses household production and consumption data obtained from the FAO's Rural Income Generating Activities (RIGA) database, combined with national price data to simulate the welfare effects of increases in staple prices. The analysis considers the impacts of changes in the prices of food staples on the revenues and expenditures of rural households in nine developing countries, and finds that most rural households are net buyers of staples and stand to lose from higher staple prices in the short run. However, simulations of the 2007/2008 food price crisis also suggest that the magnitude and timing of the welfare shocks depended heavily on the type of crops produced and consumed by each rural household.

The analysis ignores behavioural responses by households and second round effects such as changes in rural wages. Chapter 3 examines the short-to medium-term impacts of alternative agricultural policies using a behavioural model, the Development Policy Evaluation Model (DEVPEM), which admits such adjustments. The aim of the model is to provide illustrative results that show how structural diversity among developing countries, and systemic differences between developed and developing countries, can affect the welfare and distributional outcomes of alternative agricultural policy interventions. Models are constructed for six countries; two in African (Ghana and Malawi), two in Asia (Bangladesh and Viet Nam) and two in Latin America (Guatemala and Nicaragua).

The basic model takes into account several specificities of developing countries that are likely to affect the welfare impacts of agricultural policy interventions. One is the joint role of the farm household as a producer and consumer of food crops. This means that the effects of policies such as farm price support depend on what happens on both the supply side and the demand side. A second factor is that many farm households confront high transaction costs when selling output or purchasing inputs. In the extreme, these transaction costs may be so high that the farmer remains cut-off from the market altogether, producing only for home consumption (that is, subsistence). Under these circumstances a subsistence farm household may not benefit from higher farm prices, and could in fact lose via induced increases in land rental rates or in the prices paid for purchased inputs. A third aspect is that rural households are heterogeneous in terms of their income sources, expenditure patterns and ownership of factors (particularly land), and will therefore be affected diversely by the direct and indirect impacts of policies. Accordingly, DEVPEM considers the behaviour of structurally diverse agents, including commercial farms, semi-subsistence and subsistence farms, and landless rural households.

The first conclusion from these policy experiments is that no untargeted agricultural policy intervention is pro-poor within the rural economy. Market price support for food crops harms net buyers of food, often the poorest farm and non-farm (landless) households, although the proportion of net buyers varies significantly across countries. Support for cash crops does not have this drawback; however, cash crops are typically grown by farmers with relatively high incomes, so support seldom reaches the incomes of the poorest. While all agricultural policy instruments are less efficient than a simple cash payment (in terms of net rural benefits relative to urban cost), the degree of inefficiency of some market interventions, notably input subsidies, is not inevitably as high as observed in developed OECD countries. The reason is that farmers are less likely to rent land than in OECD countries and they tend to purchase fewer inputs, which in turn imply fewer leakages away from the farm level.

In general, the results show that direct payments are the most efficient way of boosting incomes in the short term, while public investments, which should also have broader long-term pay-offs, have short- to medium-term impacts that are pro-poor. These results are therefore fully consistent with the best practice advice established for OECD countries of using social policies to safeguard incomes in the short term, and non-distorting public investments to boost competitiveness in the long term. Whether there are caveats to this advice depends on the degree to which other factors germane to policy design and implementation are sufficiently important to offset the shortcomings of market interventions on both efficiency and equity criteria. These issues are taken up in Chapter 4 which considers whether there is a role for government in stabilising markets, and Chapter 5 which considers the arguments for and against using input subsidies. Market stabilisation policies were widely adopted in response to the 2007-08 food price crisis, while input subsidies, notably for seed and fertiliser, have been proposed as a possible solution to the weak performance of African agriculture, with the World Bank advocating the use of "smart subsidies" (*e.g.* for seeds and fertiliser) as a way of "jump-starting" agricultural markets (World Bank, 2007).

Price stabilisation provides an operationally simpler way of mitigating the impacts of price shocks on the poor than do market-based forms of risk management or the provision of income safety nets. But, as noted above, managing prices is a blunt instrument for affecting the incomes of the poor, since among the poor there are typically both buyers and sellers of food. Over the longer term, stable prices may promote a more predictable investment climate, but they may also deter private risk management and export instability onto world markets. A further problem is that efforts to stabilise price often turn into system support (or implicit taxation) according to the political weight of vested interests, with the result that they eventually become unsustainable from a budgetary point of view. Government interventions can also exacerbate instability if they are mis-timed, or if they induce strategic behaviour by traders (such as delaying imports in the anticipation that tariffs will be waived).

In assessing the need or otherwise for market stabilisation, it is important to identify the origin of risk (international or domestic), the degree of exposure, and the nature of the consequences. The role for national policies will depend partly on the extent to which price volatility is contained at the international level. World price changes may be transmitted onto domestic markets more fully in some countries than others, and more in some years than in others. Domestic shocks, stemming chiefly from production shortfalls, are typically more frequent than international shocks, so market openness may help reduce the frequency of shocks. But such a policy may not be sufficient to contain rare but severe international shocks. The worst case scenario is one where domestic and international shocks reinforce each other, for example when the domestic harvest fails and the government needs to purchase large amounts of imports, and there is a price bubble on the world market. The priority under these circumstances is to ensure that poor countries are provided with the finance to address this rare but potentially catastrophic scenario.

Input subsidies similarly look like a convenient mechanism for redressing multiple failings such as the under-development of infrastructure, missing markets for credit and inputs, and a lack of knowledge of the benefits of using improved seed and fertiliser. They are likewise used to address the social objective of transferring income to poor farmers. But they distort the allocation of resources and, unless carefully designed, risk impeding the development of private input markets. Their benefits can be similarly "captured" by vested interests, outliving their original rationale and becoming a drain on public finances. Innovative design features (so-called "smart subsidies") can overcome some of these problems over the short to medium term, but in the long-term there is a fundamental inconsistency in using a single instrument to address multiple market failure (*i.e.* long-term development) and social objectives. If the objective is to address market failures, then there needs to be a point at which that task is accomplished and an exit strategy is implemented. This in turn means that a system of social protection needs to be put in place and be fully operational once the subsidy is withdrawn.

For both market stabilisation and input subsidies, the general conclusion is that the benefits and costs of intervention need to be judged relative to the benefits and costs of tackling the underlying market failures directly and using social policies to address immediate concerns over income levels. While there may be circumstances under which market interventions are warranted in the short to medium term, fundamentally they treat the symptoms rather than the causes of low incomes. If such instruments are to be used, it is important that they do not crowd out essential investments in building more productive and resilient rural economies, or in developing effective economy-wide systems of social protection.

References

Collier, P. and S. Dercon (2009), "African agriculture in 50 years: Smallholders in a rapidly changing world?", FAO Expert Meeting on How to Feed the World in 2050, Food and Agriculture Organisation of the United Nations, Rome.

Hazell, P., C. Poulton, S. Wiggins and A. Dorward (2007), "The future of small farms for poverty reduction and growth", *2020 Discussion Paper,* No. 42, International Food Policy Research Institute, Washington, DC.

Morris, M., H.P. Binswanger-Mkhize, D. Byerlee (2009), *Awakening Africa's sleeping giant: prospects for commercial agriculture in the guinea savannah zone and beyond*, World Bank, Washington DC.

OECD (2001), *Market Effects of Crop Support Measures*, OECD, Paris.

OECD (2002), *Agricultural Policies in OECD Countries: A Positive Reform Agenda*, OECD, Paris.

OECD (2003), *Farm Household Income: Issues and Policy Responses*, OECD, Paris.

OECD-FAO (2011), *Agricultural Outlook 2011-2020*, OECD, Paris.

World Bank (2007), *Agriculture for Development*, *World Development Report*, World Bank, Washington, DC.

Zezza, A., B. Davis *et al.* (2008), "The impact of rising food prices on the poor", unpublished manuscript, Food and Agriculture Organisation, Rome.

Chapter 1

A Strategic Framework for Strengthening Rural Incomes in Developing Countries

by

Jonathan Brooks[1]

This chapter examines the role of agricultural policies in raising rural incomes in developing countries. The underlying premise is that policies need to be effective given current economic structures, yet anticipate and facilitate the transition to structures that are capable of generating higher incomes. This means improving the productivity and competitiveness of smallholder farmers, who dominate developing country agriculture, while widening opportunities outside the sector as the economy diversifies. Many of the required policies are not agriculture-specific, and agricultural policies need to be framed in the broader context of rural development strategies. The primary need is for investments in public goods that can support agricultural and rural development, such as agricultural research and rural infrastructure.

1.1. Introduction

This chapter motivates and proposes a strategic framework for policies to strengthen rural incomes in developing countries, and then discusses the role of different policy instruments within that framework. Raising rural incomes is central to the near-term target of the first Millennium Development Goal (MDG1), which calls for the eradication of extreme poverty and hunger, with specific targets that include halving between 1990 and 2015 the proportion of people living on less than a dollar a day. Income growth is essential for sustained poverty reduction in developing countries, although the distribution of income (and income growth) matters too. And while food security has many facets, most of the world's hungry are chronically hungry and they are chronically hungry because they are poor. Thus, higher incomes are needed in order to meet both poverty and hunger targets.

The world as a whole may achieve MDG1, thanks primarily to rapid income growth in East Asia (and China in particular). But in many parts of the world, progress has been weak or non-existent. Using a recently updated income benchmark of USD 1.25 per day, the World Bank estimates that the number of poor is actually increasing in Africa and South Asia (Chen and Ravallion, 2008). Between 1981 and 2005, the incidence of poverty in South Asia fell from almost 60% to 40%, but because of population growth that was not enough to bring down the numbers of poor. In Sub-Saharan Africa, the situation was even worse: the incidence of poverty was virtually unchanged between 1981 and 2005, at around 50%, which implied that the number of poor almost doubled from 214 million to over 390 million. By this measure, Africa's share of the world's poor increased from 11% in 1981 to 28% in 2005. Some modest signs of progress in recent years (since 2000) were arrested by the recent increase in world food prices, which the World Bank estimated was severe enough to throw another 100 million people into poverty (Ivanic and Martin, 2008). Prices have since fallen back, but remain considerably higher than they were in the first part of the decade.

If broader based progress on MDG1 is to be achieved, then average incomes will need to increase much more rapidly in the next few years than they have done in the past twenty. Given that three quarters of the world's dollar a day poor live in rural areas (corresponding to 880 million people), and most depend on agriculture for their livelihoods, there is a particular need for faster development of *rural* incomes. This, in turn, requires carefully thought out agricultural and rural development policies, and a specific consideration of what to do about smallholders, who form the backbone of developing country agriculture.

For the best part of 30 years, agriculture has been discriminated against by both national policy makers and donors. Developing countries showed an overall tendency to tax their farmers, both in terms of pricing policies (Anderson *et al.*, 2008) and through urban bias in the allocation of expenditures (Bezemer and Headey, 2008). Similarly, official development assistance to the sector declined in both absolute terms and as a proportion of total allocations, with a fall from USD 8 billion in 1980, equal to 17% of total aid, to a little over USD 3 billion in 2005, corresponding to a share of less than 4%. One reason for this "agro-scepticism" was low rates of perceived success compared with investments in other areas such as education and health (Easterly, 2008). Another was the combination of declining real agricultural prices and, in successfully developing economies, a falling share of agriculture in GDP and employment. These were interpreted by policy makers as signs of higher returns from investing in other sectors.

Timmer (2005) suggests that, in the poorest countries, such reasoning confused cause and effect. This is because agricultural investment was and is necessary to elicit the productivity gains that initiate the agricultural transformation (described later), which involves the release of resources from the sector and – when part of a balanced development strategy – their more productive allocation to other sectors. Hence, the relative "decline" of agriculture is a consequence of development success.

In recent years, policy thinking has come full circle. In the 1960s, policies centred on subsidies for seeds and fertiliser, and heavy investment in R&D and extension services, while in the 1970s, the emphasis shifted to integrated rural development (dealing with all the necessary inputs at once). These ideas were abandoned in favour of structural adjustment and "getting prices right" in the 1980s and 1990s (Adelman, 1999). Since the millennium, there has been a renewed focus on R&D, and in fixing market failures (specifically in input markets). The new thinking was reflected in the World Bank's 2008 *World Development Report*, which not only called for greater agricultural investment, but also advocated the use of "market smart" subsidies for seeds and fertiliser (*i.e.* targeted, time-bound subsidies that are operated through private markets).

The current paradigm, with faster agricultural and rural development seen as a pre-requisite for deeper economy-wide development, has been reflected in a range of commitments. In the case of Africa, for example, the African Union's 2003 Common African Agricultural Development Programme (CAADP) framework sets a target of 6% for agricultural growth, while under the

2003 Maputo Declaration its members are committed to allocate at least 10% of public expenditure to agriculture and rural development. On the donor side, the G8 pledged in 2008 to provide EUR 1 billion of support for investment in African agriculture, and in 2009 increased that commitment to EUR 15 billion over three years. At the Africa Fertiliser Summit in Abuja in 2006, African Union Member States resolved "to increase the level of use of fertiliser from the current average of 8 kilograms per hectare to an average of at least 50 kilograms per hectare by 2015".

The food price crisis of 2007-08 also triggered international commitments to allocate more money to agriculture. The World Food Programme appealed for additional funds and had received more than USD 1 billion by the end of 2008; the World Bank launched a Global Food Crisis Response Program in mid-2008; and the FAO presented an Initiative on Soaring Food Prices in June 2008 (Abbott, 2009). Bilateral donor countries also pledged additional resources to address problems in developing countries stemming from the food crisis (GDPRD, 2009). As food prices fell back, the legacy of the crisis drew attention to the deeper need for short, mid and long-term measures to tackle food insecurity and poverty. The UN High-Level Task Force on the Food Security Crisis advocated a two pronged approach, focusing on emergency relief and renewed efforts to invest in agricultural development, with a particular emphasis on supporting smallholder agriculture. With prices now lower (albeit still above average levels over the past ten years), the emphasis has shifted to the chronic lack of smallholder development, with strong support for "smart" subsidies for seed and fertiliser.

The purpose of this chapter is to give consideration to what constitutes an effective strategy for boosting agricultural and rural incomes, and to discuss the appropriate role for agricultural policy with respect to that objective. Of course, income growth (and the related objectives of poverty reduction and improved food security) is not the only objective in developing countries. Policies to improve incomes therefore need to be seen in the light of their impact on other objectives, such as sustainable resource use and improved gender equality. Furthermore, it is helpful to make a distinction between policies that can raise incomes in the short term, when economic structures are essentially fixed, and those that can facilitate a transition to economic structures that are capable of generating fundamentally higher incomes in the longer term. A fundamental thesis is that while there are some instruments that can be beneficial irrespective of the time horizon, there are nevertheless difficult trade-offs between short and long-term priorities, and a strategic framework needs to acknowledge those trade-offs.

A key focus is on policies towards smallholders, who underpin the agricultural economies of poor countries, although consideration is also given to other agriculture-dependent households (such as wage earners). The term "smallholder" refers not to farm size *per se*, but to producers with limited resource endowments relative to other farmers in the sector. Smallholders may be efficient in the technical or allocative sense, but they often struggle to be competitive, either because of their lack of endowments, or because they confront missing or under-developed markets. Insufficient farm size may also be an issue, although other assets, such as farm management skills may also be lacking.[2] The competitiveness of smallholders may also be undermined by external factors, such as subsidised exports arriving on internal markets.

The structure of this chapter is as follows. Section 1.2 describes the broad experience across countries of agricultural development and structural change, and discusses the factors that have driven those changes. This helps provide macro-level context on the evolving role of agriculture in the economy, and points to some principles that can help guide policy formulation. Section 1.3 contrasts those principles with actual agricultural policies and approaches to sectoral development. Section 1.4 proposes a strategic framework for strengthening rural incomes which seeks to reconcile the short run objective of poverty alleviation with the long-run aim of facilitating development. Section 1.5 presents some conclusions and identifies some specific priorities for analysis.

1.2. Economic development and the structural transformation

The process of economic development is characterised by three historical tendencies that are of relevance when considering the strategic options for smallholder development. These are the *sectoral* transition away from an economic structure based on agriculture to a more diversified one dominated by manufactures and services, the *spatial* tendency towards increased urbanisation, and an *institutional* transformation from an economy based largely on informal rules to one based on formal legislation (Jonasson, 2009). Other changes also accompany economic development, such as falling death rates and – with a lag – declining birth rates. In its 2008 *World Development Report*, the World Bank makes a useful distinction between agriculture-based, transforming and urbanised economies that captures the first two elements of this economic transformation.[3] Countries tend to move through these categories, although they may experience short cuts in the process, such as learning from policy experiences in developed economies, or take detours due to such factors as civil war, corrupt government or a misallocation of public resources. Moreover, the trajectory (and resulting composition of income growth) is not the same for each country, but will depend on the rate and composition of demand growth, as well as the sources and composition of productivity changes, which are partly determined by policies. There is also a strong element of historical path dependence, and the possibility that certain facets of the adjustment process may be changing systemically over time.

Before considering what drives these changes, and the appropriate role for government policy in shaping and responding to the structural transformation, we first describe in more detail the nature of adjustments that have occurred across countries, and take note of where different countries are now.

The sectoral transformation

The sectoral changes typically associated with economic development are evident from the relationship between agriculture's share of GDP and GDP per capita, which can be seen both from a cross section of countries at a given point in time as well as by looking at the evolution in individual countries over time. Figure 1.1 provides a scatter plot of these two indicators for 180 countries in 2005. For countries with a GDP per capita of USD 2 000 or less (approximately 7.5 on the logarithmic scale), it is still not uncommon for agriculture to constitute 30% or more of the economy. As per capita income rises above USD 10 000, practically no country has an agricultural sector that accounts for more than 10% of GDP.

Figure 1.1. **Share of agriculture in GDP and per capita GDP**

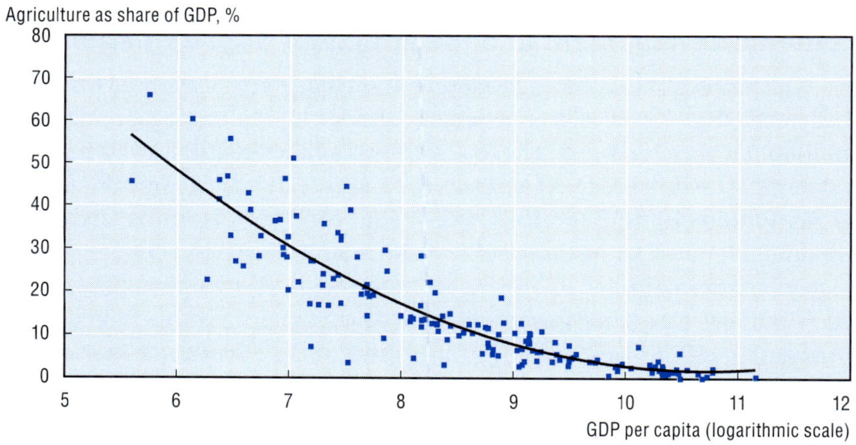

Note: 180 countries; GDP per capita refers to 2005 PPP USD.
Source: *World Development Indicators* (2009).

StatLink http://dx.doi.org/10.1787/888932552304

For a narrower group of emerging and developed economies, Figure 1.2 shows how agriculture's share of GDP changed between 1961 and 2008, with countries ordered according to their GDP per capita. The graph re-emphasises the strong inverse correlation between agriculture's share of GDP and GDP per capita, with high income OECD countries typically having no more than 2%-3% of GDP generated by their farm sectors. A second, and consistent,

Figure 1.2. **Evolution of agriculture's share of GDP in various countries (1961 to 2008)**

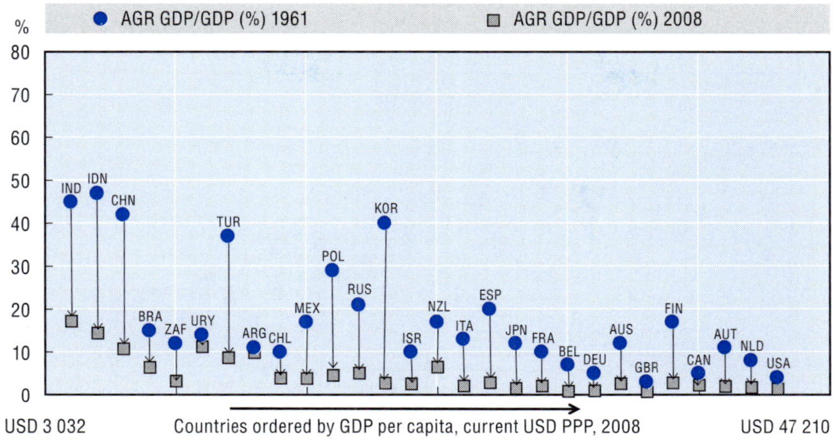

Note: Data for agriculture's share in GDP are for 2007 instead of 2008 in the cases of Israel, Japan and the United States, and for 2004 in the cases of Canada and New Zealand.

The statistical data for Israel are supplied by and under the responsibility of the relevant Israeli authorities. The use of such data by the OECD is without prejudice to the status of the Golan Heights, East Jerusalem and Israeli settlements in the West Bank under the terms of international law.

Source: FAO (1999); WDI (2008); IMF, 2008, national sources.

StatLink http://dx.doi.org/10.1787/888932552323

feature is that agriculture's share of GDP has declined in all countries, including those with a strong comparative advantage in agricultural activities. A third point is that the decline of the share of resources in agriculture has been larger for countries with lower incomes, which have more scope for agricultural productivity improvements and for shifting resources into new non-farm activities (in developed countries, that shift has already occurred).[4]

We can also look at more recent changes across three continents. Figure 1.3 shows the declines in agriculture's share of GDP since 1980 in selected African, Asian and Latin American countries (Panels A, B and C respectively). In Latin America, agriculture's share of GDP has declined to less than 10% of the total in most countries, and is less than 20% in nearly all countries. In Asia, the ratio is above 30% in a few countries but has declined to between 10% and 20% across most of the continent. Africa stands out as an exception. First, because of the significant number of countries in which agriculture still accounts for a third or more of GDP; and second because in several of the poorest countries, all with a per capita income of less than USD 1 000 per year, agriculture's share of GDP has actually *risen* over the past 30 years. The majority of these countries have had weak or negative economic growth and have been marked by conflict or civil strife, attesting to the particular development challenges faced in the region.

Figure 1.3. **Evolution of agriculture's share of GDP in Africa, Asia and Latin America, 1980-82 to 2006-08**

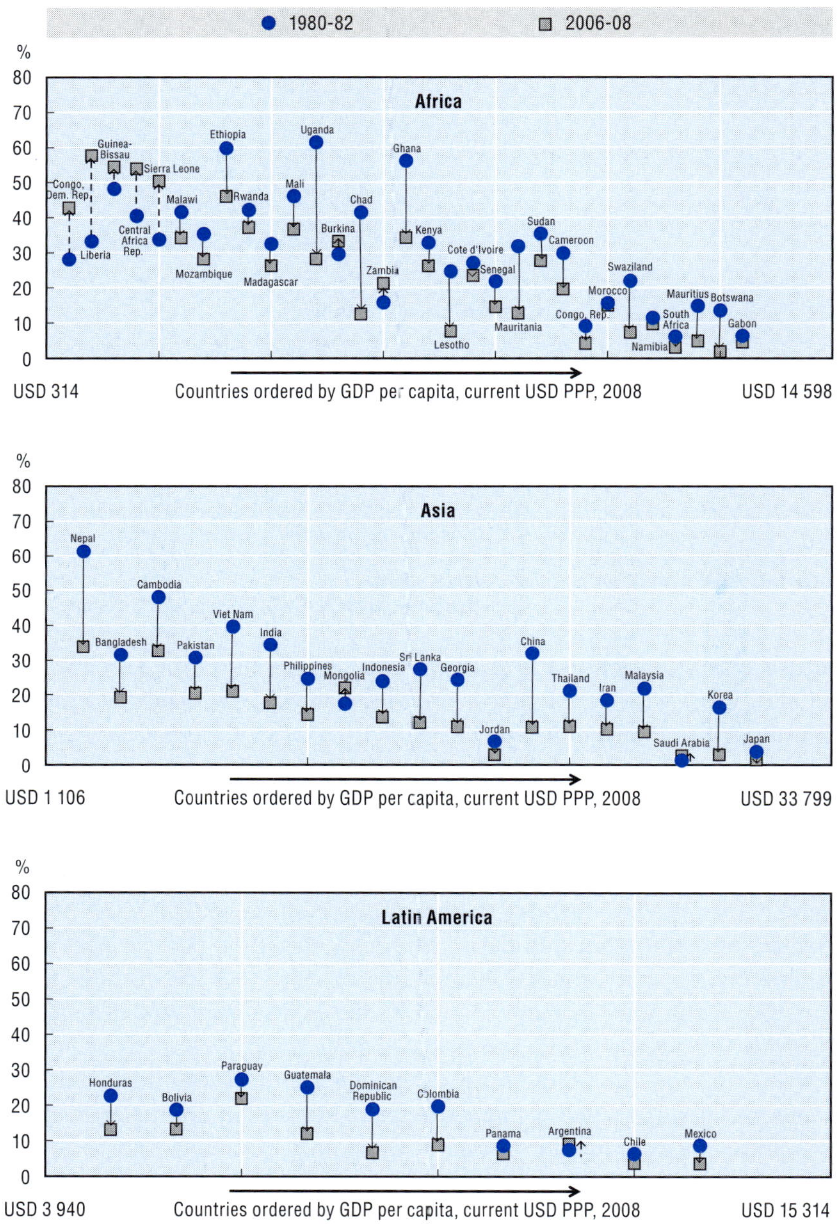

Note: 1981-82 average instead of 1980-82 for Ethiopia; 2006-07 instead of 2006-08 for Mali, Mauritania and Cameroon; 2006 instead of 2006-08 for Burkina Faso; 1993-95 average instead of 1980-82 for Cambodia; 1985-87 instead of 1980-82 for Viet Nam; 1981-82 instead of 1980-82 for Mongolia; 2006-2007 instead of 2006-08 for Iran, Malaysia and Japan.

Source: World Bank (2010), *World Development Indicators.*

StatLink http://dx.doi.org/10.1787/888932552342

The declining share of agriculture in GDP is accompanied by a release of labour to other sectors. Again, this can be seen both in cross-section and over time. Figure 1.4 shows the average shares of employment in agriculture, manufacturing, and services for 120 countries, divided into seven income categories. On average, half of the labour force in the poorest countries is occupied in agriculture, whereas for countries that have a per-capita income of USD 15 000 or higher, the service sector generally occupies two-thirds or more of the labour force, manufacturing most of the remainder, and agriculture just a few per cent.

Figure 1.4. **Employment shares in agriculture, manufacturing, and services, 2005**

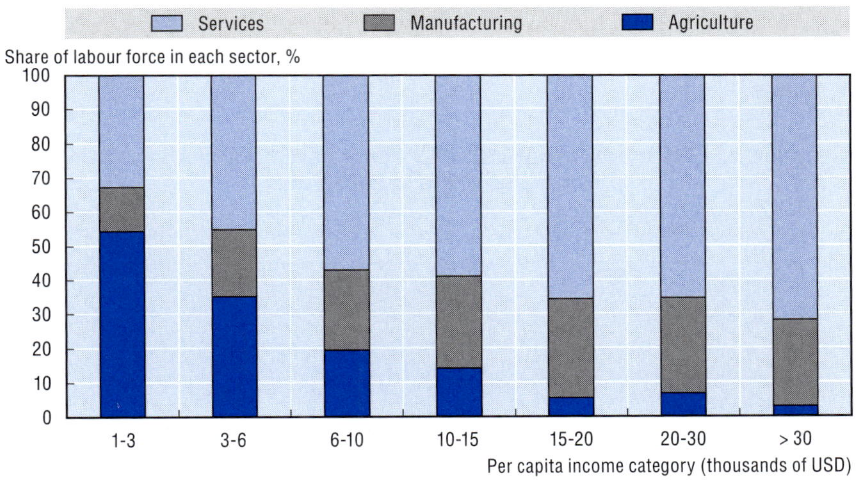

Note: 120 countries; income categories are based on GDP per capita 2005, PPP USD.

Source: *World Development Indicators*, 2009.

StatLink http://dx.doi.org/10.1787/888932552361

Changes in agriculture's share of employment are shown in Figure 1.5 (for the same countries and same period as in Figure 1.2). For most middle income (transforming) countries, the decline in agriculture's share of employment has been more rapid than the fall in the sector's share of GDP, reflecting stronger gains in labour productivity in agriculture than in other sectors, and more rapid increases in labour demand outside agriculture (Johnson, 2000). The labour adjustment has been larger for upper-middle income countries than for lower-middle income countries such as India, as alternative employment possibilities have become more widely available and the transition of labour out of semi-subsistence farming is more fully underway. In the poorest agriculture-dependent economies, where agriculture accounts for a third or more of GDP, agriculture typically accounts for two-thirds or more of employment and the release of labour from the sector has been limited (World Bank, 2008).

The general pattern seems to be that, in the early stages of development, non-agricultural growth outpaces agricultural growth, but labour is not readily absorbed by the non-farm economy. This phase corresponds to widening productivity gap between labour in agriculture and labour in other sectors. Agriculture's share of GDP thus falls more rapidly than its share of employment, a development which corresponds to increasing income inequality between the non-farm and farm sectors. At somewhat higher levels of income, however, the income gap precipitates a more rapid outflow of labour from farming and agriculture's share of employment declines more rapidly than its share of GDP. At this stage the difference in labour productivity narrows. This delayed adjustment of labour, followed by a rapid catch-up, is apparent from the historical experiences of transforming and urbanised economies.

Figure 1.5. **Evolution of agriculture's share of employment in various countries (1961 to 2007/2008)**

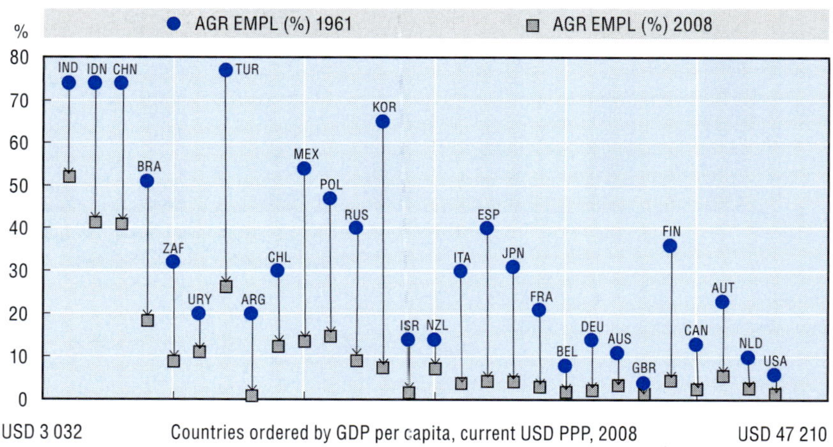

Note: 2007 or 2008 data are used according to availability for all countries except Argentina, where 2006 data are used.

The statistical data for Israel are supplied by and under the responsibility of the relevant Israeli authorities. The use of such data by the OECD is without prejudice to the status of the Golan Heights, East Jerusalem and Israeli settlements in the West Bank under the terms of international law.

Source: FAO (1999); WDI (2008); IMF (2008); national sources.

StatLink http://dx.doi.org/10.1787/888932552380

The sectoral transformation is influenced by the fact that the productivity and skills gap between subsistence agriculture and other manufacturing and service sectors is becoming ever larger. Econometric estimates suggest that "the economic growth process, as manifested in the structural transformation, has become progressively less successful at integrating low productivity agricultural labour into the rest of the economy" (Timmer, 2010). It may indeed be more difficult for farm labour in poor agriculture-dependent economies to be absorbed by other sectors than it was for, say, European farmers to move into industrial jobs a century earlier. Yet once the sectoral transformation is underway, its pace is invariably more rapid than in the past (Table 1.1). Whereas it took a century or more for agriculture's share of GDP to fall from 40% to 7% in OECD countries that went through the industrial revolution early, middle income countries are effecting these changes in three decades or less. This accelerating change is matched by an even more rapid release of labour out of the sector. In Korea, agriculture's share of employment fell from 40% to 16% in just 14 years – a transition which took 53 years in the United States and 68 years in the United Kingdom (the first country to go through the industrial revolution).

Table 1.1. **Pace of adjustment in various countries, based on agriculture share of GDP and employment**

	Agriculture share of GDP			Agriculture share of employment		
	Year of 40%	Year of 7%	Years required	Year of 40%	Year of 16%	Years required
Netherlands	1800	1965	165	1855	1957	102
Denmark	1850	1969	119	1920	1962	42
United Kingdom	1788	1901	113	1800	1868	68
Chile	1875	1980	105	1950	1993	43
Mexico	1890	1992	102	1969	2000	31
United States	1854	1950	96	1897	1950	53
France	1878	1972	94	1921	1965	44
Brazil	1910	2003	93	1960	2005 (20.5%)	> 45
Germany	1866	1958	92	1900	1942	42
Japan	1896	1969	73	1940	1971	31
Poland	1935	1991	56	1968	2006 (18.7%)	> 31
India	1962	2006 (17.5%)	> 44	2005 (58%)		–
China	1967	2006 (11.7%)	> 39	2006 (43%)		–
Turkey	1970	2007 (8.9%)	> 37	1998	2007 (28.7%)	> 9
Korea	1965	1991	26	1977	1991	14
Indonesia	1971	1997	26	2006 (42%)		–

Source: Adapted from Kim, H. and Y.K. Lee (2003).

While there are common elements of the adjustment process, there are also asymmetries. Figure 1.6 compares agriculture's share of GDP with agriculture's share of employment for recent years in a large number of developed and developing countries. The 45-degree line corresponds to a situation in which the sector's share of GDP and its share of employment are equal, implying that labour is as productive in agriculture as in other sectors.[5] Nearly all countries lie below this line, suggesting that agricultural labour is

Figure 1.6. **Agriculture's share of GDP *versus* agriculture's share of employment, 2008**

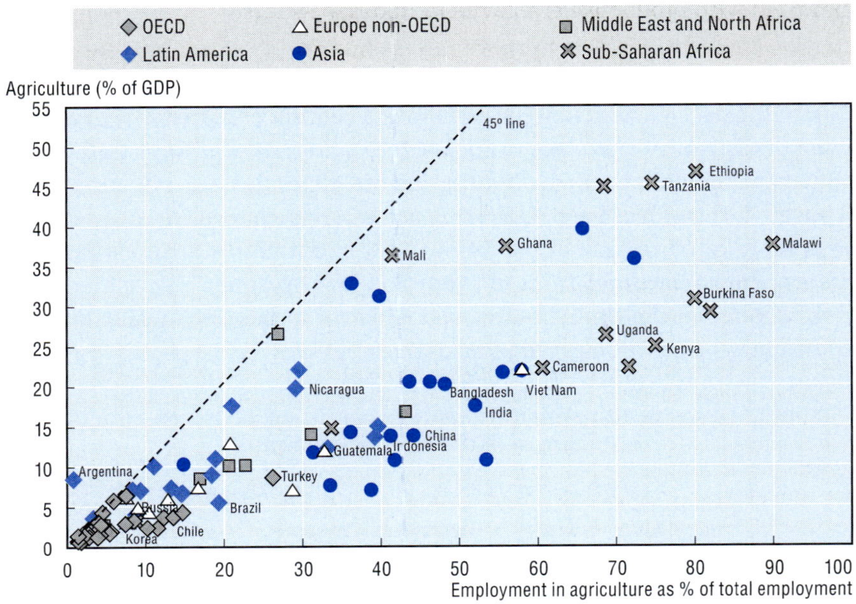

Source: World Development Indicators (2010) and national sources.

StatLink http://dx.doi.org/10.1787/888932552399

less productive than non-agricultural labour, but there are wide differences across countries. As economies develop, and agriculture's share of both income and employment declines, labour productivity in agriculture tends to converge with that in other sectors. But in some developing countries, that convergence appears to be deferred. This has important implications in terms of the distribution of income across the economy.

In a large number of developing countries, agriculture's share of employment is more than twice its share of GDP. Even allowing for off-farm income, this implies that labour generates much less value added in agriculture than it does in other sectors. However, there are some developing countries in which labour's implied productivity in agriculture is closer to its productivity in other sectors, suggesting that there is no pre-ordained trajectory for the sectoral transformation. Improvements in the productivity of agricultural labour will by definition tend to close the gap directly, while the outflow of less productive labour to more productive occupations in other sectors will narrow it indirectly. The countries furthest from the 45-degree are those in which agricultural productivity has languished and the sector has not been effectively assimilated into the rest of the economy. In practice, both forms of integration need to go hand-in-hand: for most countries improvements in agricultural labour productivity will cause labour to be released from agriculture (increased production is unlikely to be fully absorbed by higher domestic demand or exports), and that labour needs to be provided with productive employment in other sectors.

Nevertheless, the challenges of managing the sectoral transformation in poor African economies such as Burkina Faso, Ethiopia and Malawi, where agriculture dominates employment and accounts for a third or more of GDP, are likely to be different to those in Asian economies such as Bangladesh, India and Viet Nam, where agriculture still accounts for about a half of employment but the sector's share of GDP has dropped to 20% or less.

The spatial transformation

The spatial transformation from rural to predominantly urban economic activity is not as uniform as the sectoral transformation. Figure 1.7 shows that a majority of countries with a per capita income of less than USD 5 000 (approximately 8.5 on the logarithmic scale) have more than 50% of their population in rural areas. On average, this share declines to 25% when countries reach an income of USD 20 000. Urbanisation may occur both as a result of higher birth rates in urban areas compared to rural areas and as a result of rural-to-urban migration.[6] In China a majority (about 56%) of the population is still rural, but rapid migration from rural areas might soon change this situation. In 1983 the cumulative number of rural migrants was about 2 million in China. This number had increased to about 78 million in the year 2000. Six years later, in 2006, the estimated cumulative number of rural migrants was 132 million (OECD, 2009). In India, which has the largest rural population in the world (approximately 800 million), it is estimated that rural-to-urban migration accounts for about 30% of urbanisation (Mitra and Murayama, 2008). In Brazil, rapid migration from rural areas increased the share of the population in urban areas from 15% in 1940 to 56% in 1970, and to more than 80% in 2000 (Wagner and Ward, 1980; Brazilian Demographic Census 2000).

The agglomeration of human activity with economic development may be inevitable, but the specific character of urbanisation is not. Rural areas may become more "urbanised" as a result of the agricultural transformation and the induced growth in non-farm activity. Alternatively, urbanisation may result from poor unskilled labour migrating to cities in the anticipation of improved prospects. In the absence of a parallel development in social infrastructure, the resulting shanty towns (common in Asia and Latin America) may impose severe social strains. Hence there is a need to plan for a sustainable form of spatial development, which is likely to require the balanced promotion of farm and non-farm opportunities in rural areas.

Figure 1.7. **Share of population that is rural and GDP per capita (190 countries)**

Note: 2007 figures for 190 countries; GDP per capita measured in 2005 PPP USD.

Source: *World Development Indicators* (2009).

StatLink http://dx.doi.org/10.1787/888932552418

As labour has left farming for other sectors, we nevertheless see a tendency for poverty to remain predominantly rural. Three-quarters of the world's dollar a day rural poor (nearly 600 million people) live in "transforming" economies, where poverty remains predominantly rural but agriculture contributes less than 20% to overall economic growth (World Bank, 2008). This imbalance points to the need for an integrated *rural* strategy, not just an agricultural one.

The institutional transformation

A third dimension of change that developing countries tend to go through is the institutional transformation from an economy based largely on informal rules and procedures to one based on a framework of formal legislation – in short, the transformation from informal to formal institutions.[7] In the absence of formal rules that effectively regulate employment, property ownership, or land use, various types of informal rules and procedures are usually applied instead. Two examples from the agrarian economy are sharecropping as a means to overcome moral hazard situations in the farmer-labourer relation and "squatter's rights", which regulate access to land.

Informal institutions may be efficient in the context of existing economic structures, but a strengthening of formal institutions may facilitate the emergence of more commercially oriented agriculture, for example by supporting the development of land rental and credit markets, and other forms of formal contracts. Formal institutions may also make the use of certain policy instruments more feasible, such as social safety nets. It is thus the third element of development strategy, which involves facilitating the three dimensions of structural change: adjustment away from agriculture-dependence, a desirable form of urbanisation and concomitant institutional development. The specifics of this strategy are taken up further in this chapter.

Promoting a pro-poor structural transformation

Within a broad strategy based on facilitating structural change across its three dimensions, are there ways in which policy makers can promote a more "pro-poor" pattern of development?

Several studies have suggested that, in poor countries, this can be achieved most effectively by agricultural growth, with a specific focus on smallholder development (for example, Hazell *et al.*, 2007; FAO and the World Bank, 2009). The direct impacts of smallholder growth are likely to be pro-poor because the majority of the poor live in rural areas, where agriculture is the dominant economic activity and smallholder farming is the predominant structure. Smallholder development can increase returns to assets that the poor possess – their labour and in some cases their land – and push down the price of staples, which is crucial when so many of the poor are net buyers of food. Indirectly, the benefits of smallholder growth are also likely to be pro-poor, because of potentially strong linkage effects to the rest of the economy. Empirically, a number of studies have confirmed that agricultural growth tends to be effective in reducing poverty (Irz *et al.*, 2001; de Janvry and Sadoulet, 2009).

With development, one would expect the economy to diversify and for growth outside agriculture to become progressively more important as a source of income growth, and hence of poverty reduction and improved food security.[8] Moreover, in the long-term manufactures and services are capable of generating annual growth rates of 10% or more, whereas growth rates in agriculture seldom exceed 5%. How then should governments strike the balance between boosting incomes in the context of existing structures, where there is more immediate scope for poverty reduction, and facilitating the transition to a more diversified economy which has the potential to generate much higher average incomes? In order to address this question, it is helpful to look more closely at the mechanics of the sectoral transformation, in particular the structural changes within agriculture that accompany the inter-sectoral shifts described previously.

In aggregate terms, the sectoral transformation is driven by a range of supply and demand interactions. At the outset of the development process, productivity growth in agriculture leads to the generation of surpluses that induce a demand for other goods and services. On the demand side, once basic needs are met, income elasticities of demand for food tend to be less than for other consumption, so the demand for food grows more slowly than the demand for other goods. Longer term changes on the supply side are less clear. There has been an historical tendency to assume that labour productivity grows less rapidly in agriculture than in the manufacturing sector, partly because there is less scope for benefiting from the division of labour. Indeed dual models of transition, such as the Lewis Model (Lewis, 1954) typically contrast a stagnant traditional rural sector with a dynamic and modern manufacturing sector. However, this view has been challenged by others, who have observed rapid science-based technical change in agriculture (for example, Hayami and Ruttan, 1985). For a range of low and middle income countries, Martin and Mitra (2001) found that, on average, total factor productivity gains in agriculture outpaced those in other sectors. However, the average masks a range of different experiences. For example, agricultural productivity grew faster than productivity in manufactures in India, Kenya and Turkey, but slower in Egypt, Korea and Indonesia. Which effect dominates depends on a range of factors, including national institutions and government policies (for example, towards innovation).

When agricultural productivity growth outpaces productivity growth in other sectors, we tend to observe a more rapid release of labour – the "push" out of the sector complementing the "pull" from relatively fast demand growth outside the sector (but possibly within the rural economy). Despite this release of resources, the agricultural sector typically continues to expand in absolute terms. Those leaving the sector tend to be those who have not participated to the full extent in productivity improvements, and have thus seen their profitability undermined, and those who have superior prospects in other sectors. On the other hand, if agricultural productivity growth is relatively slow, then that transition occurs more slowly and – when non-agricultural demand growth is weak – may not occur at all.

The pace at which resources are released from agriculture also depends on the international forces of supply and demand. Rising net demand for exports from overseas may slow the adjustment process, while increasing import competition in the form of lower prices may accelerate it. The balance of these pressures should, in principle, be a reflection of a country's comparative advantage. Historically the terms of trade have tended to move against agriculture, as supply growth has outpaced demand growth. Insofar as prices reflect production costs, declining real prices are not a problem for innovative farmers, whose productivity changes are responsible the price changes. However, for inefficient farmers, there is clearly a threat to profitability which, along with the possibility of improved opportunities in other sectors, can determine exit decisions.

Timmer (1998) has suggested that these basic supply and demand forces are associated with a four-phase adjustment process for the agricultural sector. In the early stages of development, agriculture dominates output and employment, and the priority is to "get agriculture moving". The subsequent generation of a surplus within agriculture leads to a second period in which agriculture makes a key contribution to growth both directly and via a variety of linkages. In the third phase, agriculture's share of national income declines and agricultural incomes fall behind those in other sectors – often inducing political tensions. The fourth and final phase is one in which the agricultural sector, including agricultural labour markets, are integrated into the rest of the economy.[9] This pattern is consistent with a non-linear adjustment process where, initially, employment in agriculture declines more slowly than the sector's share of GDP, but then there is a more rapid release of labour from the sector as average incomes improve.[10]

A number of poor countries, mostly in Africa, are at the first two stages of this development process. In its 2008 *World Development Report*, the World Bank identifies approximately 170 million rural people in agriculture-dependent economies (mostly in Sub-Saharan Africa) falling below a USD 1.08 a day income threshold in 2002 (measured in 1993 purchasing power parity dollars). A much larger number of rural poor – 583 million in 2002 – live in transforming economies, a large proportion of them in China and India. The majority of the rural poor in Latin America live in urbanised countries. In transforming and urbanised economies, where agriculture accounts for less than 20% of overall growth, there is a general need to integrate agriculture more fully with thriving parts of the economy.

These different circumstances suggest a need for differentiated strategies, with agriculture prioritised at early stages of development.[11] In terms of establishing the framework conditions, this is likely to imply a greater role for policy in overcoming market failures and a greater share of public expenditures going to providing essential public goods (such as agricultural research and rural infrastructure). It also points to the need for caution in generalising from the experiences of a few agriculture-dependent countries. For example, Malawi's input subsidy programme has received much attention, but it should be remembered that Malawi is amongst the poorest countries in the world, with agriculture's share of GDP more than one-third and poverty predominantly rural (nearly 90%). With market failures likely to be more endemic in poorer countries, it has been suggested that a different set of agricultural policy instruments is appropriate – a point taken up below. A key point to note, however, is that even in the majority of what the World Bank characterises as agriculture-dependent economies, non-agricultural activities still account for the majority of national income, and cannot be ignored as a source of employment opportunities.

Furthermore, even in agriculture-dependent economies, the process of structural change within agriculture is likely to lead to a shedding of farm labour. There has been much debate over the relative efficiencies of small versus large farms. A range of benefits from small scale family farming have been noted. For example, farm labour may be easier to motivate and supervise, while smallholders may have important local knowledge and may be more adept at managing some forms of risk. On the other hand, there are important economies of scale beyond the farm in areas such as procuring inputs, obtaining information on markets and technical farming issues, in meeting standards and certifying production, and in transacting with large scale buyers from processors and supermarkets, with their exacting demands (Wiggins, 2009).

Some of this debate may be beside the point, insofar as the relevant comparison is between small farms of less than a few hectares and not between small farms of less than a hectare and mega farms of several hundred hectares or more. Small farms may be technically and allocatively efficient, given the existing level of development, but economies of scale beyond the farm gate are likely become more apparent as the economy develops. In much of Africa, only a minority of farms produces a marketed surplus. A change in average farm size from say 0.5 ha to 2 ha, with more farmers recording market sales, would still correspond to small scale farming, but would nevertheless require a significant degree of farm level adjustment – in particular the release of labour from the sector. Moreover, a progressive consolidation would appear to be necessary to generate substantially higher per capita incomes, irrespective of efficiency considerations. An FAO and World Bank study (FAO and World Bank, 2009) has suggested that in Africa's Guinea Savannah zone, an unexploited area that shares many similarities with the Brazilian Cerrado and the Northeast Region of Thailand, opportunities abound for Africa's farmers to compete effectively in regional and global markets. Improving opportunities for farmers within agriculture are likely to be associated with output rising but labour leaving the sector – as has been apparent from the agricultural transformation in most OECD countries.

1.3. Policy responses to structural change

Policy makers face a number of challenges as a consequence of the structural pressures identified in the previous section. Fundamentally, the only way of generating a sustainable increase in real incomes is through raising productivity. In order to redress the gap between urban and rural incomes, that implies addressing the underlying problem of lagging labour productivity in agriculture.

In poor countries still in the early stages of economic development (*i.e.* in agriculture-dependent economies), the paramount need is to boost average incomes in general and rural incomes in particular. In these economies most of the poor live in rural areas, so raising rural incomes tends to have the most immediate impact on poverty. Furthermore, insufficient income is the principal cause of food insecurity, which is also more prevalent in rural areas. In the long term, however, the ultimate need is to shift from improving incomes in current activities to facilitating adjustment into activities that command potentially higher incomes. The balance to be struck, therefore, is between policies that support development of the agricultural sector in general, but do not prevent structural adjustment from occurring as a direct consequence of that process. In general, spending on public goods, such as infrastructure and agricultural research, can facilitate both developments, whereas direct subsidies to farm activities – notwithstanding potential benefits that are discussed later – run the risk of impeding adjustment.

Historically, poor countries have tended to tax their agricultural sectors rather than subsidise them (Krueger, Schiff and Valdés, 1991). They have done this both explicitly, for example via export taxes and regulated food prices, and implicitly, by investing relatively less in rural areas. Since the mid-1980s, there has been a reduction in this tendency, but it still prevails (Anderson *et al.*, 2008). As incomes rise and as agriculture's share of employment decreases, countries find they can afford more easily to provide support to their agricultural sectors and the opposite tendency is observed. In the 1990s, developing countries on balance made that switch, with their average nominal rate of assistance (NRA), which measures the degree to which domestic farm gate prices are above adjusted border prices, becoming positive (Figure 1.8).[12] On average, they thereby joined high income OECD countries in providing protection to their agricultural sectors. Note that this measure only takes account of relative prices, and ignores subsidies to farmers or other aspects of agricultural spending. Also, some caution needs to be exercised in interpreting these numbers, as the NRAs are weighted averages for import-competing products, exportables and non-tradables, and in some cases different patterns can be observed when these categories are treated separately.[13]

By implication, the level of protection is also linked to agriculture's share of employment. The horizontal axis of Figure 1.9 shows agriculture's share of total employment, while the vertical axis measures the NRA. The arrows show the movement for each country between 1961 and 2005. For nearly all developing countries, the arrow points to the north-west, indicating an increasing rate of protection as labour leaves the sector, whereas the pattern for high income OECD countries is mixed.[14] The arrows are also much longer for developing countries, as more dramatic structural changes have taken place, and the associated change in protection has been larger. Interestingly,

Figure 1.8. **Nominal rate of assistance to agriculture in developed and developing countries, 1955-2004**

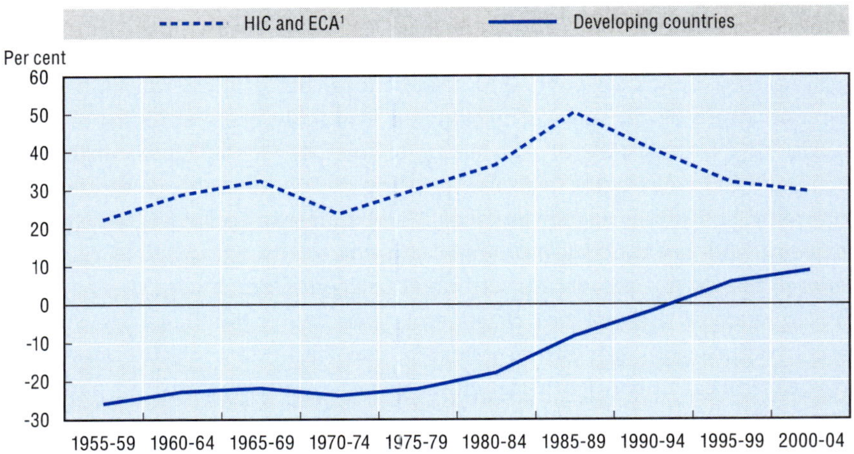

1. HIC = High income countries; ECA = Eastern European and Central Asia countries.
Source: Anderson *et al.* (2008).

StatLink http://dx.doi.org/10.1787/888932552437

Figure 1.9. **NRA and agriculture's share of employment, 1961 and 2005, selected countries**

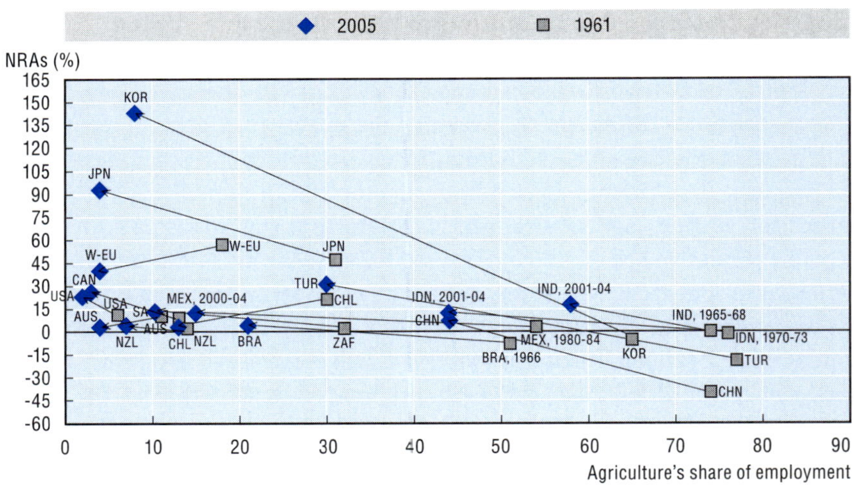

Source: Anderson *et al.* (2008); FAO (1999); WDI (2008).

StatLink http://dx.doi.org/10.1787/888932552456

developing countries have undergone significant adjustment, seemingly irrespective of whether the rise in protection has been large or small.

Programmes targeted at smallholders are common in emerging (and developed) economies. For example, both Brazil and Chile have programmes that seek to integrate smallholders into the commercial sector, notably via the use of subsidised credit and investments in farm-level infrastructure. In few cases, however, have policy makers openly acknowledged that long-term competitiveness is not a realistic goal for the majority of smallholders and decided to focus their programmes on potentially viable operations. At the same time, there is no documented case of a smallholder programme in which the majority of farmers enrolling have succeeded in progressing through the programme to successfully join the ranks of efficient commercial producers. In other words, no programme has reversed the structural tendency for smallholders to leave the sector. This suggests that these policies constitute social policies at least as much as developmental ones.

In a number of developing countries, the movement away from dis-protection has been matched by commitments to allocate more resources to agricultural development. A large number of developing countries with relatively neutral pricing policies are at a critical juncture in terms of agricultural policy development. Do they spend scarce resources on supporting farmers directly, or do they invest in the broader underpinnings of agricultural development and economic development more generally?

In part, this question turns on how policy makers choose to address the issue of smallholder adjustment. Smallholders in developing countries often underpin the rural economy, yet they face systematic adjustment pressures as a necessary corollary of the development process. With technology improving, and more efficient use being made of scarce resources, including the exploitation of scale economies, smallholders that do not participate in sectoral cost improvements inevitably face pressure on their incomes. Governments can shield smallholders from this pressure, or they can help them adapt to it – either by becoming more competitive, obtaining incomes from other sources, or by finding jobs outside the sector.

In choosing the appropriate policy mix, the preceding discussion suggests a need for a logical framework that acknowledges three important things. First, economic structures are relatively fixed in the short to medium term, so efforts to achieve an immediate impact on incomes need to be based on an understanding of how people currently earn their livelihoods. Second, over the long-term, structures change and the sectoral transformation implies that the inter-generational future for the majority of smallholders cannot lie exclusively in farming; hence there is a need for policies that enhance households' opportunities outside the sector as well as within it. In other words, agricultural policies are not enough. Third, in order to improve both agricultural competitiveness and the prospects for earning more outside the sector, the most important policies may not in fact be agricultural policies. It is therefore important that smallholder policies are framed in an economy-wide context, with agricultural policies a component of the overall policy mix. The elements of an appropriate strategy are discussed in the next section.

1.4. A strategic framework for strengthening rural incomes and facilitating smallholder adjustment

In proposing a strategic framework for strengthening rural incomes it is helpful to make a distinction between the short to medium term issue of how best to support incomes, reduce poverty and tackle food insecurity (beyond immediate questions of humanitarian relief), and deeper long-term questions regarding how best to consolidate those gains via broader economic development. There may be a connection, with programmes that are effective in the short-term sowing the seeds for longer term development, but there may equally be trade-offs, so it is conceptually helpful to distinguish short-term social objectives from long-term development ones.

At the same time, a long-term strategy for development needs to acknowledge the inevitability of the sectoral, spatial and institutional transformations that accompany economic development. This means acknowledging the inevitability of structural change and the need for policies that smooth adjustment across each of its dimensions. The strategy proposed below consists of: *i)* enabling smallholders to become competitive or boost their incomes from other sources (value addition, diversification or exit); *ii)* promoting a broader rural development strategy that does not focus exclusively on agricultural development, but seeks to create a more diversified rural economy; and *iii)* strengthening institutions with a view to reducing the need for second best instruments.

The proposed framework suggests potential roles for government policy in providing the optimal balance of opportunities both within and outside agriculture. Following a discussion of the ability of alternative policy instruments to address short-term and long-term objectives, we illustrate the positive role that government can play with a case study of Thailand, which has successfully transformed from an agrarian country to an urbanised economy based around manufacturing (Box 1.1). Rural poverty has fallen dramatically, as have hunger and malnutrition, while agricultural production has increased and a new class of commercially viable farmers has emerged.

Box 1.1. **Thailand's agriculture: Transition and sustained growth**

The agricultural transformation

Since 1960 Thailand has transformed from agrarian country to an urbanised economy based around manufacturing and services. Agricultural output has grown more quickly than population, with two distinct periods of growth.

From the early 1960s to the early 1980s, agriculture grew by about 4% per year. It did so by opening up new lands for farming, and using more labour, to

Box 1.1. **Thailand's agriculture: Transition and sustained growth** *(cont.)*

produce more of the main staples for both domestic market and for export. During this period, agriculture was a main driver of the economy. In 1980, more than 70% of the active population and most of the country's poor was employed in agriculture. The sector produced cheap food for the growing urban population and was a major export earner, with rice exports prominent. This was achieved despite taxation of the sector to fuel industrial growth. Rice exports were taxed directly and exporters were further penalised indirectly by over-valuation of the baht.

From the mid-1980s, agriculture began to transform. Labour left agriculture, attracted by jobs in manufacturing, urban services and the rural non-farm economy. At the same time, the land frontier was closing and it became harder to add new land. Consequently, agricultural growth slowed to about 2%-3% per year, although productivity increased notably. Given opportunities in both domestic and international markets, new activities emerged such as rubber, cassava, pineapples, and high value perishables for the fast-growing cities. Most farm households have diversified their income sources, while some have become more specialised in higher-value agricultural products sold into more sophisticated marketing chains. The rural non-farm economy has grown to the point where it provides around half of all rural jobs. Successful industrialisation of the country has allowed direct and indirect net taxation of farming to be virtually eliminated.

In the early 1960s more than 60% of the rural population lived in poverty. By the early 2000s that had been cut to only a little more than 10%. From 1988 to 2007, the number of households affected by food poverty declined from 2.55 million to 418 000. With more and cheaper staples, and reduced poverty, child malnutrition has also declined. The incidence of underweight young children fell from 17% in 1987 to 7% in 2006; while that of stunting was reduced from 25% to 16%. During the 1960s and 1970s, most of the improvements came from increasing farm incomes. Subsequently, incomes from rural non-farm jobs and remittances from migrants have become important. Not only has poverty fallen, but the reductions have been similar across the provinces.

The role of government policy

The Thai story is an example of a successful transition from an initial situation in which it was possible for agriculture to grow by putting underused factors of production to work, with only limited improvements in productivity, to a later stage where land and labour have become increasingly scarce and growth could only continue through improved returns to these scarce factors. The figure below shows how, from the late 1980s onwards, land and labour have started to leave farming, while productivity growth of these two factors has accelerated.

Box 1.1. **Thailand's agriculture: Transition and sustained growth** *(cont.)*

Figure 1.10. **Land, labour use and productivity in Thai agriculture, 1961 to 2007**

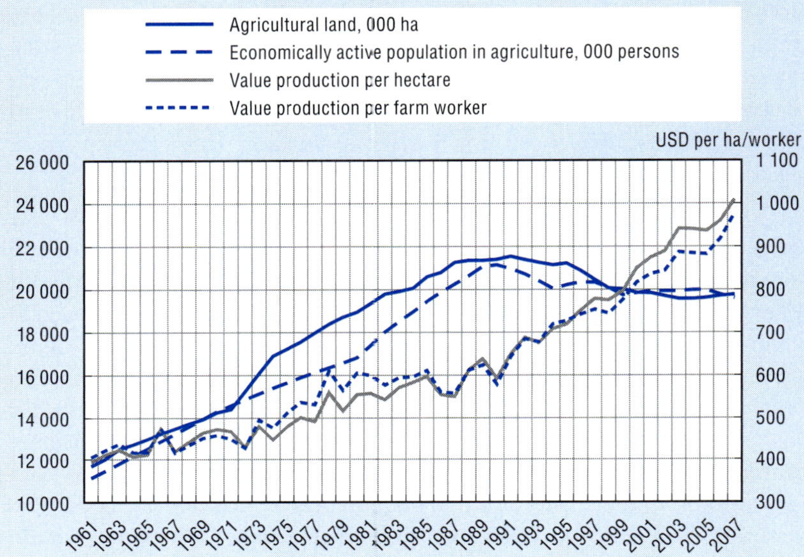

Source and notes: Computed from FAOSTAT data. Agricultural land and economically active population in agriculture, left scale; value of production per hectare and per worker, right scale. Labour force in agriculture taken as the estimated economically active population in agriculture.

StatLink http://dx.doi.org/10.1787/888932552247

Success has been achieved primarily through private initiative, with the state playing a strategic role in setting an investment climate, investing in roads and research, and also supporting agricultural credit to overcome market failures. Specifically:

- The opening of new land was encouraged by tenure rules that allowed family farmers to clear enough land, 4 to 8 hectares, for a small farm and then gain secure property rights so long as they paid taxes.
- The state built roads that provided access to new lands and constructed irrigation works that were especially important before 1980 or so. Thereafter these investments were complemented by more spending on rural education, electrification, and telecommunications. Public investment in large-scale irrigation gave way to policies supporting private investment in smaller-scale irrigation.

Box 1.1. **Thailand's agriculture: Transition and sustained growth** *(cont.)*

- Public agricultural research has contributed as well. With intensification of Thai agriculture, uptake of agriculture research outputs has increased; as seen with modern rice varieties, use of improved rubber trees, and improved varieties for maize, soybean and cassava.
- To resolve failing rural credit markets, the Bank of Thailand instructed all commercial banks to allocate 5% of all commercial loans for agriculture at an interest rate lower than the market. The Bank for Agriculture and Agricultural Cooperatives (BAAC) has subsequently expanded provision of agricultural credit to 90% of farm households and all farm co-operatives, using a group liability guarantee which enables small farmers to access short-term credit without land title deeds as collateral.
- More recently the state has promoted certification to allow farmers to develop premium domestic markets;
- Government has sought to promote agricultural exports through active participation in trade talks, in multilateral, regional and bilateral negotiations.
- Although agriculture was heavily taxed in the early stages of industrialisation, the state had the wisdom to ease this subsequently, once manufacturing was not so dependent on farming for resources.

Source: Material provided to OECD by Leturque and Wiggins (2010).

Short-to-medium-term policy considerations

The optimal way of addressing short-term social objectives is with social policies. Across a range of developed and developing countries, population-wide social safety nets have been used to support the incomes of rural households. In developing countries, conditional cash transfers (CCTs) have become particularly popular over the past decade. These programmes transfer cash to generally poor households on the condition that they make pre-specified investments in the human capital of their children. CCTs have been found to be effective at increasing consumption levels among the poor and have led to behavioural changes, although their impact on *final* outcomes in health and education has been less clear (Fiszbein and Schady, 2009). This may be due to the need for CCTs to operate in conjunction with complementary investments (*e.g.* in schools and hospitals). An issue with CCTs is when the "conditional" element is warranted. For example, it may not be worth incurring the monitoring and enforcement costs associated with the condition that parents put their children in school if they would do that anyway.

A virtue of cash transfers is that they do not distort production and consumption decisions and thereby lead to the kinds of efficiency losses associated with market interventions. From a political economy standpoint, however, they appear to be "welfare", whereas market interventions are often justified on other grounds (even if they are disguised welfare). In low income countries, where family and social networks are the main form of social protection, there is some wariness about potentially weakening that aspect of the social fabric.

In the poorest countries, however, it has been argued that the necessary institutions and infrastructure do not exist for cash-based instruments to be appropriate, and agricultural policies, such as price supports and input subsidies, have been suggested as an alternative. For example, there may be no registry or information base by which to establish criteria of eligibility; remote farmers may not have a convenient way of spending cash; and – with weak institutions – such programmes may be particularly susceptible to corruption.

For these reasons, market interventions such as price supports and input subsidies have been proposed as a more practical way of raising the incomes of farm households. In the case of OECD countries, the Policy Evaluation Model (PEM) has shown such instruments to be inefficient, because a large share of the benefits leaks to non-farm factor owners (principally landlords) and suppliers of purchased inputs (OECD, 2001). Moreover, the use of such instruments typically has perverse distributional effects, with larger farmers benefiting more than small ones (OECD, 2003). For developing countries, the effectiveness of such instruments relative to direct payments is currently being investigated with a new model, the Development Policy Evaluation Model (DEVPEM), which adapts the PEM to take account of some specific aspects of developing country agriculture (OECD, 2011a). One aspect of this new model is the joint role of the farm household as a producer and consumer of food. As a consequence of this joint role, one would expect price support to be particularly ineffective in developing countries. Indeed, amongst the poor, there are both net buyers and net sellers of food, and many (perhaps the majority of) farm households may in fact be net buyers.

On the other hand price stabilisation has been proposed as way of containing the impact of adverse shocks on producers and consumers. The difficulties of designing price stabilisation programmes are well known. Formal (*ex ante*) price stabilisation induces moral hazard, with agents failing to mitigate risk, and price stabilisation can easily turn into systemic price support or suppression, depending on political pressures. Price stabilisation programmes have also proven to be costly and often financially unsustainable (Anderson and Roumasset, 1996). Yet, in the short term, it has been suggested that there may be no other way of containing the impact of adverse price shocks on poverty and food security than by seeking to offset those impacts directly (for example by releasing/buying stocks, or by changing tariffs). The advantages and disadvantages of alternative approaches to market stabilisation are discussed in (OECD, 2010a).

Input subsidies have also been suggested as a way (possibly the only way) of targeting the incomes of poor farmers, with the attraction (when markets are insulated) of lowering prices to consumers too. A host of difficulties of using input subsidies have been acknowledged. The inevitability of leakages to other agents in the supply chain, and the difficulties of targeting have already been noted. In addition such measures may crowd out the development of private input markets, may lead to the over-use of inputs, and once introduced have historically proven difficult to rescind. Nevertheless, there has been renewed optimism that a new generation of so-called "smart" subsidies, by virtue of innovative design features, such as exit strategies, can deliver income benefits while limiting their known shortcomings (Dorward, 2009b). These arguments are considered in (OECD, 2010b).

An additional (and sometimes dominant) argument that has been used for input subsidies, and to a lesser extent for price support, is that it acts as a bridge to longer term development, creating a surplus among farmers that can initiate the agricultural transformation described earlier. This argument, a justification for the policy focus on smallholder development, starts from the premise that economic development has to start with improving the profitability of existing structures. From a conceptual point of view, this argument needs to be kept separate from the short-term social rationale for intervention. Over the longer term, policy makers need to consider *why* farmers are not competitive. This may be because of high transaction costs, for example due to poorly developed road systems, or market failures, such as the absence of functioning credit markets. The optimal policy solution would be to reduce transaction costs, via suitable investments and thereby correct market failures directly – in other words, treat the causes of a lack of competitiveness rather than the symptoms. However, such structural policies can take time to pay-off, so direct support for smallholder development (with an emphasis on input subsidies) has similarly been advanced as a practical alternative. In the case of the poorest of economies, this may be part of a much broader package of specific help to improve farmers' competitiveness. Ideally, long-term development policies should be able to discriminate between those who are potentially viable in the sector and those who are not, creating improved competitive conditions for the former and facilitating adjustment via diversification or exit for the latter. Generalised agricultural policies, such as price support or sector-wide input subsidies, cannot do this. Indeed they run the risk of impeding structural adjustment.

Long-term priorities

Elements of a long-term strategy for improving farm household incomes, with a focus on smallholder adjustment, are set out in Table 1.2. Smallholder adjustment here is understood to be the optimal path to higher long-term income, be that improved competitiveness within the sector, income diversification (from agricultural or non-agricultural sources), or exit to other sectors. Adjustment pathways are described in the columns, and policy instruments in the rows. The first column (improving competitiveness within agriculture) applies to farm households only, but the other columns may apply to both farm households and salaried (often "landless") worker households. Note that the adjustment pathways (columns) are not mutually exclusive: for example, one household member can enhance the farm's competitiveness while another provides off-farm income. Also, the instruments (rows) do not exhaust all possible policies, but focus on those with persuasive arguments.

In respect of farm households, it is important to have a realistic view of which farmers have the potential to succeed commercially within the sector. In some regions agro-ecological conditions may be such that farming may not be inherently commercially viable. More generally, the appropriate adjustment pathway may depend on the basic type of farming system. For example, in East and Southern Africa the scope for agricultural growth in areas where a mixed maize and cash crop system dominates is inherently stronger than the potential in areas where rainfed sorghum and millet combine with pastoral agriculture (Dixon *et al.*, 2001). Yet even when agro-ecological conditions are inherently favourable, the nature of structural change is such that farm operations tend to consolidate into fewer and more efficient enterprises, and some farmers leave the sector.[15]

Given the need to acknowledge that some farmers will succeed while others will not, and the impossibility of identifying exactly which farmers fall into each category, the main role for policy would appear to be in providing public goods that can improve competitiveness, but impose few distortions to incentives at the margin, such as investments in rural infrastructure, skills and training, and R&D.[16] Such investments are unlikely to crowd out the development of other activities and potential income streams, although they are likely to accelerate the pace at which more efficient operators absorb and replace less efficient ones. Most of the relevant expenditures would need to be made at the economy-wide or sectoral level rather than in the form of payments to individuals. A further role for policy is when there are endemic market failures, for example in credit markets. Access to credit is important for smallholders, and private credit markets may find it not worth their while to engage with smallholders, simply because of their size and the difficulties of becoming informed about the creditworthiness of many small operations.

Table 1.2. **Strategic framework for strengthening farm household incomes**

Policy instrument	Adjustment pathway				
	Help farmers become more competitive within agriculture	Diversify income sources		Leave the sector for off farm work	Social protection for those unable to adjust
		Within agriculture	Outside agriculture		
Price policies	Treats symptoms of uncompetitiveness rather than causes	May impede adjustment			Price stabilisation proposed as a second best safety net
Input subsidies	Treats symptoms of uncompetitiveness rather than causes		May impede adjustment		Proposed as second best instrument for the poorest countries
Credit policies	May correct market failures	Indirect impacts			
Investment in human capital	Minor effects of formal education for this generation; technical training more appropriate for productivity.	Can help farm family members and rural workers move into skilled jobs	Important for farm family members and rural workers	Important for managing inter-generation change	
Investment in infrastructure	Helps with market integration	Helps improve local job opportunities		Can ease migration decisions for offspring	
R&D and extension	Public and private sector important; gains from adoption and adaptive research.	Can expand agricultural employment			
Labour market reforms		Important for raising employment opportunities and wage incomes			
Income transfers (possibly conditional cash transfers)				Conditional school attendance may complement investments in schools	Preferred policy for those unable to adjust
Regional policies	Important for improving market integration	Expanded non-farm activity would raise farm wages	Important for building a diversified rural economy with wider job opportunities		
Develop producer associations	Reduce transaction costs and help exploit economies of scale	Indirect impacts			
Land policies and property rights	Need to encourage rental markets and facilitate land purchases by small farmers			Secure property rights and rental markets can ease exit decisions	

In many developing countries, farmers may have insecure land rights, while land rights rental markets function poorly or do not exist at all. Secure land rights can improve incentives for investment in the land, and can also facilitate the development of rental markets. The latter can in turn help compensate for market failures, provide flexible responses to economic and productive incentives, allow farmers to invest in farming capital, and help the poor and young gain access to land under conditions that are less demanding than those required to participate in land sales markets. Renting land may also be a first step to future land acquisition. The under-development of rental markets may prevent the consolidation of land into more productive units, thus impeding agricultural investment and making it more difficult for uncompetitive farmers to diversify out of the sector.

Improving the competitiveness of farm households

Income diversification for farm households and salaried agricultural workers

Income diversification is essential for many farm households. For the poorest farm households, this is likely to provide some insurance and is in effect a "coping" strategy. For other farm households, having one or more family members draw income from outside agriculture may be the start of a successful move into more remunerative activities. Policies that support farm income alone, such as market price support, act as a disincentive for income diversification outside agriculture, and create an obstacle to one of the key "adjustment pathways". The key policies required to help households diversify their income sources are again those that improve human capital. Regional development policies, including the development of rural infrastructure, may also have an important role.

Leaving the sector for skilled employment

Ultimately, the majority of smallholders in developing countries, or at least their descendants, will have stronger prospects outside the agricultural sector than within it. The most important need, if not for this generation then for the next, would therefore appear to be investment in the education and skills that would enable households to command higher wages. At the same time, well-defined property rights, especially with respect to land, are important for farmers to be able to cash in their assets, and exit the sector on favourable terms.

Regional development programmes, by targeting economic assistance to less developed regions, may also have a role in bringing jobs to people (rather than the other way round) and so can prevent the problems associated with mass migration into cities. However, rural policies are not fundamentally agricultural policies (nor *vice versa*). Regional policies can boost development within and outside agriculture, but without biasing household decisions about how best to invest for the future.

In many middle-income countries, the conditions of salaried agricultural work are at least as important as the development of small scale farm entrepreneurs. In Chile, for example, two-thirds of all households receiving the majority of their income from agricultural sources are salaried workers, not farmers. Labour market policies have an important role in ensuring that core standards of employment are met, while improved labour market flexibility has been suggested as a way of reducing informality (OECD, 2008).

Social protection for households that cannot adjust

Many poor households, notably older ones, face severe limitations in their adjustment potential, irrespective of the policies that are in place (for example, resource poor and post retirement age farmers). Hence, the need for social protection to address chronic as well as transient income shortfalls. Investments in human capital (notably education) and measures such as contingent cash transfers can ensure that the next generation makes a quantum leap in terms of development.

1.5. Conclusions

This chapter has pointed to the inevitability of structural change in the agricultural and rural economy, the consequent implications for adjustment among smallholders and other agriculture-dependent households, and the associated need for policies that facilitate rather than impede that process. A strategic framework has been proposed to assist policy makers in choosing the appropriate mix of policy instruments.

A key premise of that framework is that, for the majority of agriculture-dependent households, their long term (*i.e.* inter-generational) future lies outside the sector. This is true even in the context of improving opportunities within the agriculture sector as a whole. Hence, while policies need to improve opportunities within agriculture for those who potentially have a competitive future in the sector, they also need to create wider opportunities for who do not, or could earn higher incomes elsewhere. For both types of development path, many of the necessary policies will not be agriculture-specific, so it is important that agricultural policies are framed in a broader economy-wide framework, and that rural development policies are not equated with agricultural policies.

Justifications for the use of agricultural market interventions (either in output or input markets) rely on the notion that they are a pragmatic option when ideal choices are not available: market interventions induce efficiency losses and hence do not provide a theoretically optimal way of providing social protection (where social safety nets are to be preferred); nor, in the absence of market failure, are they the ideal way of fostering growth, since they treat the symptoms of a lack of development rather than its underlying causes. By contrast, the provision of public goods (including investment in agricultural research) is not just theoretically superior but of proven value. Here, there is a case for prioritising agriculture at early stages of development and gradually shifting the balance of public spending as the economy develops and diversifies.

Nevertheless, plausible reasons have been advanced for why, given weak institutions, high transactions costs and endemic market failures, some agriculture-specific interventions might be desirable. For such arguments to be properly substantiated, there needs to be a clear distinction between instruments that are effective at raising incomes and reducing poverty in the short-term, and those that have an enduring effect by promoting long-term economic development. There may be trade-offs as well as complementarities between the two. For example, input subsidies may have an immediate impact on the incomes of the poor, but could thwart the development of private dealer networks and ultimately impede agricultural development. Similarly, price stabilisation may provide a quick response to extreme price shocks, but can be difficult to sustain systemically over the long term. Arguments concerning the use of these two particular policies are explored in Chapter 4 and Chapter 5 respectively. One particular issue which needs to be investigated is the short to medium term welfare and distributional impacts of alternative policies. These are investigated with a rural economy-wide model, the Development Policy Evaluation Model (DEVPEM), in Chapter 3.

The strategic framework presented in this chapter seeks to order an analysis of which types of policies are most appropriate for smallholder farmers and can contribute to faster progress on MDG1. It also attempts to delineate the role and limitations of agricultural policies in particular. A central conclusion is that it is the *policy mix* that matters, so empirical analyses of policy effectiveness need to take account of possible complementarities and trade-offs between alternative agricultural and non-agricultural instruments. The former may include the complementarities between agricultural extension and the development of infrastructure and broader investments in human capital; the latter, the opportunity cost of using different expenditure mechanisms (*e.g.* providing input subsidies versus making longer term investments in rural roads or in non-agricultural areas such as health and education). A more formal analysis of these linkages would help inform policy choices.

Notes

1. The author is grateful to Dalila Cervantes-Gody for earlier collaborative work on smallholder development, which helped motivate this chapter, to Erik Jonasson and Joanna Komorowska for technical input, and to Florence Bossard and Clarisse Legendre for statistical assistance.

2. It is important to note that what constitutes a small farm may differ markedly from one country to the next. For example, the average farm size in many Asian countries is less than a hectare, whereas much larger operations in Latin America (ten hectares or more) may be considered as small.

3. According to this classification, agriculture-based economies are those in which agriculture contributes 20% or more to overall economic growth; transforming economies are those in which agriculture contributes less than 20% to total growth yet 60% or more of the country's poor live in rural areas; while urbanised economies are those in which agriculture contributes less than 20% to overall growth and less than 60% of the poor live in rural areas. More than 80% of the rural poor in Sub-Saharan Africa live in agriculture-based countries, while over 90% of the poor in Asia, the Middle-East and North Africa live in transforming economies. A majority of Latin America's poor live in urbanised countries, although nearly one-half of the poor still live in rural areas. There are virtually no countries where agriculture contributes more than 20% to growth but in which the numbers of urban poor exceed the numbers of rural poor. Among developing countries, there is a strong correspondence between these three categories and three income classes for countries (low income, lower-middle income and upper-middle income) also specified by the World Bank.

4. There are some exceptions, such as Brazil and Chile, where the changes have been large in absolute terms, but low relative to other countries at similar income levels. In these particular countries, import substitution industrialisation policies led to a rapid growth in manufacturing prior to the base year, bringing down agriculture's share of GDP; while more recently the liberalisation of policies has mitigated the tendency of resources to shift out of agriculture, as these countries have exploited their natural comparative advantage in agricultural activities.

5. However, caution should be exercised in interpreting specific numbers as the definition of agriculture used in measuring GDP may not be the same as that used in recording employment. Moreover, labour employed in agriculture also earns off-farm income.

6. Rural areas may grow "urban" if they reach the population threshold that defines an urban area. Thus, the rate of urbanisation depends to a certain degree on how urban and rural areas are defined. Usually, population agglomerations of 5 000 people constitute the lower threshold for what is officially defined as an urban area (Haggblade *et al.*, 2007).

7. Institutions are understood here as the "rules of the game" that shape and guide human behaviour (North, 1990). The distinction between formal and informal lies largely in the enforcement mechanism. While formal institutions are usually enforced by official entities (such as police, bureaucrats and courts), informal institutions are socially sanctioned norms of behaviour that rely primarily on self-enforcement mechanisms of obligation, expectations of reciprocity, and internalised norm adherence (de Soysa and Jütting, 2007).

8. Christiansen *et al.* (2010) find, from econometric analysis based on national surveys, that agriculture is significantly more effective than non-agriculture in reducing poverty among the poorest of the poor. It is also up to 3.2 times better at reducing

dollar-a-day headcount poverty in low-income and resource-rich countries (including those in sub-Saharan Africa), at least when societies are not fundamentally unequal. However, when it comes to the better-off poor (reflected in the USD 2 a day measure), non-agriculture has the edge. These results are driven by the much larger participation of poorer households in growth from agriculture and the lower poverty-reducing effect of non-agriculture in the presence of extractive industries.

9. These four phases are associated, respectively, with the work of Mosher (1966); Johnston and Mellor (1961); Schultz (1964) and Ruttan (1977); and Johnson (1973).

10. It is also consistent with Kuznet's hypothesis that income inequality first increases with economic development and then declines after a certain average income is attained.

11. Dorward (2009a) has proposed three types of transformation strategy: "hanging in" strategies, which are concerned to maintain and protect current levels of wealth and welfare in the face of threats of stresses and shocks; "stepping up" strategies, which involve investments in assets to expand the scale or productivity of existing assets and activities; and 'stepping out' strategies, with accumulation of assets to allow investments or switches into new activities and assets.

12. Here the NRA is expressed in percentage terms, *i.e.* as NRA = 100*(Px/Py – 1), where Px is the domestic farm gate price and Py is the adjusted border price.

13. Nevertheless, the same pattern is borne out by the authors' calculation of Welfare Reduction Indices (WRIs). The WRI is defined as the uniform trade tax which would generate the same reduction in national welfare as the prevailing structure of national distortions, Because the contribution of each product is related to the square of the price distortion, this measure captures the higher welfare costs associated with peak tariffs, and overcomes the problem whereby positive and negative NRAs tend to cancel each other out.

14. The tendency of countries to protect their agriculture as they become more developed stems from the political economy of structural change. On the demand side, as consumers spend a declining share of their incomes on food they become "rationally ignorant" that they are paying elevated prices for their food – it is not worth the effort of becoming informed and protesting. On the supply side, the release of labour from the sector means that a given transfer to each producer imposes a progressively smaller burden on the overall economy. Moreover, competitive pressures on less efficient farmers increase their incentive to lobby for government support.

15. Poulton and Wiggins (2005) present some evidence of declining farm sizes in developing countries, mostly for countries where the average farm size is a hectare or less. This is more likely to represent a fragmentation of operations, for example due to inheritance laws and property rights systems, than it is the relative efficiencies of small farms (*e.g.* ease of labour supervision; local knowledge) versus larger operations (knowledge of markets and technology; access to credit and inputs; ease of risk management; ability to assure quality).

16. There is evidence to suggest that improvements in agricultural productivity have a strong effect in reducing poverty (Irz *et al.*, 2001). There is also evidence that agricultural growth has helped support broader economic growth (for example, Tiffin and Irz, 2006), although agriculture's role as a necessary driver of development has been questioned (Gardner and Tsakok, 2008).

References

Abbot, P. (2009), "Development dimensions of high food prices", paper prepared for OECD, Paris.

Adelman, I. (1999), "The role of government in economic development", *Working Paper* No. 890, University of California, Berkeley.

Anderson, J. and J. Roumasset (1996), "Food insecurity and stochastic aspects of poverty", *Asian Journal of Agricultural Economics*, Vol. 2, No. 1, pp. 53-66.

Anderson, K., E. Jara, M. Kurzweil, D. Sandri and E. Valenzuela (2008), *Distortions to agricultural incentives: A global perspective, 1955 to 2007, Working Paper*, DECRG-TR, World Bank, Washington, DC.

Bezemer, D. and D. Headey (2008), "Agriculture, development, and urban bias", *World Development*, Elsevier, Vol. 36(8), pp. 1342-1364, August.

Cervantes-Godoy, D. and J. Brooks (2008), "Smallholder adjustment in middle-income countries: Issues and policy responses", *OECD Food, Agriculture and Fisheries Working Papers*, No. 12, OECD, Paris.

Chen, S. and M. Ravallion (2008), "The developing world is poorer than we thought, but no less successful in the fight against poverty", *Policy Research Working Paper Series* 4703, The World Bank.

Christiaensen, L., L. Demery and J. Kuhl (2010), "The (evolving) role of agriculture in poverty reduction: An empirical perspective", *Working Paper* No. 2010/36, UNU-WIDER World Institute for Development Economic Research, United Nations University, April.

de Janvry A. and E. Sadoulet (2010), "Agricultural growth and poverty reduction: additional evidence", *World Bank Research Observer*, Vol. 25, pp. 1-20.

de Soysa, I. and J. Jütting (2007), "Informal institutions and development: how they matter and what makes them change", in J. Jütting, D. Drechsler, S. Bartsch, and I. de Soysa (eds.), *Informal institutions: How social norms help or hinder development* (pp. 29–44), OECD Development Centre, Paris.

Dixon, J., A. Gulliver and D. Gibbon (2001), *Farming systems and poverty: improving farmers' livelihoods in a changing world*, FAO and World Bank, Rome and Washington, DC.

Dorward, A. (2009a), "Integrating contested aspirations, processes and policy: Development as hanging in, stepping up and stepping out", *Development Policy Review*, Vol. 27(2). pp. 131-146.

Dorward, A. (2009b), "Rethinking agricultural input subsidies in a changing world", paper prepared for the Food and Agriculture Organisation of the United Nations, Rome.

Easterly, W. (2008), "Planners vs searchers in African agricultural aid", *FAO Pro-Poor Livestock Policy Initiative*, FAO, Rome.

FAO (1999), *Improving the income of farmers and rural people*, FAO, Rome.

FAO and the World Bank, *Awakening Africa's sleeping giant: Prospects for commercial agriculture in the Guinea Savannah zone and beyond*, FAO, Rome and the World Bank, Washington, DC.

Fiszbein, A. and N. Schady (2009), *Conditional cash transfers: Reducing present and future poverty*, World Bank Publications, Washington, DC.

Gardner, B. and I. Tsakok (2007), "Agriculture in economic development: Primary engine of growth or chicken and egg?", *American Journal of Agricultural Economics*, Vol. 89(5), pp. 1145-1151.

GDPRD (2009), *The food price crisis and the global food security challenge*, Global Donor Platform for Rural Development, Bonn.

Haggblade, S., P. Hazell, and T. Reardon (2007), "Introduction", in S. Haggblade, P. Hazell, and T. Reardon (eds.), *Transforming the rural nonfarm economy: Opportunities and threats in the developing world* (pp. 3-24), The John Hopkins University Press, Baltimore.

Hayami, Y. and V. Ruttan (1985), *Agricultural development: An international perspective*, Johns Hopkins University Press, Baltimore.

Hazell, P., C. Poulton, S. Wiggins and A. Dorward (2007), "The future of small farms for poverty reduction and growth", *2020 Discussion Paper*; No. 42, International Food Policy Research Institute, Washington, DC.

IMF (2008), *IFS International Financial Statistics Online*, *www.imfstatistics.org/imf/logon.aspx*.

Irz, X., S. Wiggins and C. Thirtle (2001), "Agricultural productivity growth and poverty alleviation", *Development Policy Review*, Vol. 19, pp. 449-466.

Ivanic, M. and W. Martin (2008), "Implications of higher global food prices for poverty in low-income countries", *Policy Research Working Paper Series* 4594, the World Bank.

Johnson, D.G. (1973), *World agriculture in disarray*, St. Martin's Press, New York.

Johnson, D.G. (2000), "Population, food and knowledge?", *American Economic Review*, Vol. 90(1), pp. 1-14.

Johnston, B. and J. Mellor (1961), "The role of agriculture in economic development", *American Economic Review*, Vol. 51, pp. 566-593.

Jonasson, E. (2009), *Labor markets in transformation: Case studies of Latin America*, Lund Economic Studies, Number 155.

Kim, H. and Y.K. Lee (2003), "Agricultural policy reform and structural adjustment: Historical evidence from the Korean experience?", Policy Reform and Adjustment Workshop, 23-25 October 2003, Imperial College London, Wye Campus.

Krueger, A.O., M. Schiff and A. Valdés (1991), *The political economy of agricultural pricing policy*, Johns Hopkins University Press for the World Bank, Baltimore.

Lewis, A. (1954), "Economic development with unlimited supplies of labor", *Manchester School of Economic and Social Studies*, Vol. 22, pp. 139-91.

Martin, W. and D. Mitra (2001), "Productivity growth in agriculture and manufacturing?", *Economic Development and Cultural Change*, Vol. 49, pp. 403-23.

Mitra, A. and M. Murayama (2008), "Rural to urban migration: A district level analysis for India", *IDE Discussion Paper* No. 137, Institute of Developing Economies.

Mosher, A.T. (1966), *Getting agriculture moving: Essentials for development and modernisation*, Published for Agricultural Development Council by Frederick A. Praeger, New York.

North, D. (1990), *Institutions, institutional change and economic performance*, Cambridge University Press, Cambridge.

OECD (2001), *Market Effects of Crop Support Measures*, OECD, Paris.

OECD (2003), *Farm Household Income: Issues and Policy Responses*, OECD, Paris.

OECD (2008), *OECD Review of Agricultural Policies: Chile*, OECD, Paris.

OECD (2009), *Agricultural Policies in Emerging Economies*, OECD, Paris.

OECD (2010a), "Stabilisation policies in developing countries after the 2007-08 food crisis", TAD/CA/APM/WP(2010)44/FINAL, OECD, Paris.

OECD (2010b), "The use of input subsidies in developing countries", TAD/CA/APM/WP(2010)45/FINAL, OECD, Paris.

OECD (2011a), "Modelling the distributional impacts of agricultural policies in developing countries: The Development Policy Evaluation Model (DEVPEM)", *OECD Food, Agriculture and Fisheries Working Papers* No. 50, OECD, Paris.

OECD (2011b), "The Development Policy Evaluation Model (DEVPEM): Technical documentation", *OECD Food, Agriculture and Fisheries Working Papers* No. 51, OECD, Paris.

Poulton, C. and S. Wiggins (2005), "The future of small farms achieving pro-poor growth through agriculture: The challenges", presentation at Overseas Development Institute, *www.odi.org.uk/events/FutureAgricultures_2005/meeting_2dec/ifpri_files/frame.html*.

Ruttan, V. (1977), "The green revolution: Seven generalizations", *International Development Review*, Vol. 19, pp. 16-23.

Schutlz, T.W. (1964), *Transforming traditional agriculture*, Yale University Press, New Haven.

Tiffin, R. and X. Irz (2006), "Is agriculture the engine of growth?", *Agricultural Economics*, Vol. 35, pp. 79-89.

Timmer, P.C. (1998), "The agricultural transformation", in C.K. Eicher and J.M. Staatz (eds.), *International Agricultural Development*, Johns Hopkins University Press, Baltimore.

Wagner, F.E. and J.O. Ward (1980), "Urbanization and migration in Brazil", *American Journal of Economics and Sociology*, Vol. 9, pp. 249-259.

WDI (2008), *World Bank Development Indicators*, *ddp-ext.worldbank.org/ext/DDPQQ/member.do?method=getMembers&userid=1&queryId=6* .

Wiggins, S. (2009), "Can the smallholder model deliver poverty reduction and food security for a rapidly growing population in Africa?", *FAC Working Paper* No. 8, Future Agricultures Consortium.

World Bank (2007), *World Development Report* (2008), *Agriculture for Development*, Washington, DC.

Chapter 2

Distributional Impacts of Commodity Prices in Developing Countries

by
Mateusz Filipski and Katia Covarrubias[1]

In this chapter, household production and consumption data are combined with national price data to simulate the welfare effects of increases in staple prices. The focus is on the rural sectors of nine developing countries and on six types of staple crops. The results show that since most rural households are net buyers of staples, they stand to lose from higher staple prices in the short run. However, simulations of the 2007/2008 food price crisis suggest that the magnitude and timing of the welfare shocks depended heavily on the type of crops produced and consumed by each rural household. Simulations up to 2018 suggest higher future prices threatening welfare, but also creating opportunities for those who can increase staple production.

2.1. Introduction

The years 2007 and 2008 saw an unprecedented rise in food prices worldwide. Internationally traded grains provide a striking illustration: during 2008, the world export prices of maize, rice and wheat all peaked at more than three times their levels of January 2005 (FAO 2009). In that same year, the FAO food price index reached its highest levels ever, and remained over 200 from February to July (2002-04 averages serve as the 100 base). In 2009, prices fell back, but they remained above pre-crisis levels. Looking ahead, the latest *OECD-FAO Agricultural Outlook* suggests that the historical downward trend of world wheat prices will continue, but starting from higher levels. Prices of coarse grains will also eventually resume their historical downward trend, but not before 2015. Rice prices will remain high compared to the past, and rice will gradually become more expensive relative to wheat. Furthermore, the report also warns that "another major price spike like the one recently witnessed cannot be ruled out" (OECD-FAO, 2009).

Therefore, concern over staple prices in the near future stems from two facets: systematically higher levels, and greater volatility than in the past. These concerns come at a time when hunger statistics are already disquieting. The number of hungry people in the world has just crossed the one-billion line (FAO, 2009). The half-century-long declining trend of undernourishment has been reversed: the numbers and proportion of undernourished people rose not only during the crisis, but also in 2009 (FAO, 2009). To avoid a worsening of this situation, and possibly turn it around, it is essential to understand the link which exists between prices on commodity markets and world hunger. The aim of this chapter is to participate in this effort by shedding further light on the welfare and distributional impacts of price changes.

The focus of this chapter is on the rural sectors of a sample of developing countries. While the effects of high food prices are unambiguous for urban populations, they are complex and ambiguous for rural farm households, who stand to gain as producers and lose as consumers. In practice, in the rural sectors of developing countries, net sellers of staple foods are often a minority. For the nine countries studied in this report, the proportion of rural net sellers of staples is lower than that of rural net buyers, except in Viet Nam. The 2008 World Development Report, using a different set of countries, reports that a majority of the rural poor are net buyers of tradable staples (World Bank 2008, p. 109). Thus, the effects on the rural populations are far from clear cut.

Furthermore, three quarters of the world's poor live in rural areas (IFAD, 2001) and since the poor tend to spend a high proportion of their income on food (reflecting "Engel's law"), they are naturally the ones most likely to experience hardship in the face of rising food prices, and the most in need of relief policies.

Table 2.1. **Proportions of net buyers and net sellers of staple foods**

Country	Survey year	Urban net buyers	Rural net buyers	Rural net sellers
Bangladesh	2000	32	50	18
Ghana	1998	33	56	11
Guatemala	2000	42	50	8
Malawi	2004	12	82	6
Nepal	2003	28	36	36
Nicaragua	2001	53	37	10
Pakistan	2001	37	52	11
Panama	2003	51	44	5
Viet Nam	1998	27	24	49

Note: Staples considered included all values of cereals, pulses, roots and tubers and oilseeds produced or consumed. Specific crops differed by survey.
Source: RIGA datasets.

This work seeks to further an analysis performed in the summer of 2008 by an FAO team led by Alberto Zezza (2008). Their paper used the Rural Income Generating Activities (RIGA[2]) database to perform simulations of a rise of food price impacts on various groups of rural households, and to calculate the resulting income shocks at the household level. It then sought to identify the most vulnerable segments of rural populations, categorised by expenditure quintiles, by land ownership, by gender of the household head, among other criteria. Because their goal was to maximise cross-country comparability, they simulated an identical shock in all countries: a hypothetical 10% increase in staple prices (regardless of what the staples were). Written in the midst of the crisis, their work offered a highly informative stylised comparison of the impacts of rising staple prices, with an explicit focus on identifying characteristics of vulnerable households. Their analysis, however, used a composite "staple" category without differentiating between crops, and used no actual price information.

We offer two different perspectives: a retrospective look at the effects of the 2007/2008 crisis, and a prospective look into the potential implications of price developments in the next nine years. The former makes use of a disaggregated FAO-compiled dataset of prices made available online recently as part of the Global Information and Early Warning System (GIEWS[3]). The latter uses the ten-year price predictions of the AGLINK-COSIMO model.[4] Both datasets provide time series of food prices for various crops or groups of crops, at the country level (though sometimes at the regional or city level). Making use of such price datasets makes it possible to narrow the focus to specific crops, specific locations and the timing of price shocks. This generates simulations of the market pressures as they are experienced by households with more accuracy. Differentiating between crops is particularly important if we believe that some staples may become more expensive relative to others, as the Agricultural Outlook predicts for rice versus wheat. In the face of changing food prices, which countries suffer most? Which crops are responsible for the strongest pressures on welfare? Are there any "buffer" crops that help mitigate the welfare shock? Does the timing of price increases in different crops play a role? Which countries are hit first? Which suffer the longest food crisis?

The remainder of this chapter attempts to provide answers to these questions, and is organised as follows. In the next section, the methodology and data are described. In Section 2.3, we simulate a theoretical 10% shock on staple prices, which provides an illustration of the mechanisms at play and a cursory analysis. Section 2.4 then presents results obtained from the retrospective simulations of actual price shocks on various countries, using the GIEWS historical dataset of prices. Section 2.5 provides corresponding results from a prospective point of view, using the AGLINK-COSIMO price forecasts. Section 2.6 draws the conclusions.

2.2. Methodology and data

Methodology

Our methodology is based on Zezza *et al.* (2008). The goal is to estimate, at the household level, the welfare impacts that are induced by price shocks. A household's welfare shock is valued as the immediate effect of a price variation on the value of the household's incomes and expenditures. For prices of agricultural commodities, this means the welfare effect is equal to the increase in the value of the commodity a household is producing; minus the increase in the amount it spends to consume that same commodity (which is different from "economic surplus" measures). Because we assume the household does not have time to adapt the quantities it will produce or consume, this corresponds to the "immediate welfare effect" (following Zezza *et al.* [2008]). Formally, this immediate welfare effect for a given household facing an increase in a given crop price (all else being equal) is expressed as:

$$\frac{\Delta w}{x_0} = \frac{\Delta p}{p_0}(P - C) \qquad (2.1)$$

Where Δw is the first order approximation of the change in welfare of the household, and x_0 the household's income before the shock (approximated by total expenditures). *P* and *C* are, respectively, the values of the household's production and consumption, both as shares of x_0. Given the lack of reliable price data, producer and consumer prices are assumed to be equal in our analysis, and denoted by *p* (initial price is subscripted by 0).

The quantity P-C, net sales of a crop as a percentage of total expenditures, is also called the "net benefit ratio" (henceforth NBR). The immediate welfare effect thus reduces to the NBR of a given crop multiplied by the size of the price shock. This method of modelling a welfare effect is unsophisticated compared to what can be obtained from a behavioural model, such as the Development Policy Evaluation Model (DEVPEM) presented in Chapter 3. Indeed, we ignore all higher-order effects of the price shock: the household's adjustments in production and consumption patterns, the economy's responses in terms of labour markets, in terms of markets for substitutes and complements of the commodity in question, etc. Our approach, however, requires minimal data and still provides a solid base for cross-country comparisons.[5] It has been used in several studies (Deaton, 1989; Budd, 1993; Barrett and Dorosh, 1996), and more recently applied by Ivanic and Martin (2008) in a paper similar in spirit to this work, but using aggregated world prices rather than country-level prices.

Data: RIGA, GIEWS and AGLINK-COSIMO datasets

As mentioned in the introduction, this work uses three main sources of data: household survey data from RIGA, historic prices from GIEWS and prospective prices from AGLINK-COSIMO.

The RIGA database is comprised of a series of standardised household-level datasets for a range of developing countries. Each country's dataset is compiled from raw data available from the World Bank Living Standards Measurement Survey (LSMS) archive. For this project, we used household-level incomes and expenditures related to each commodity of interest. The detailed methodology, explaining the compilation of RIGA datasets, can be found in Carletto (2007). Our study focuses on nine countries: four in Asia (Bangladesh, Nepal, Pakistan, Viet Nam), two in Africa (Ghana, Malawi), and three in Central America (Guatemala, Nicaragua, Panama). These countries were selected so that various regions of the world were represented and because comparable data could be found for all of them.

The GIEWS database offers a variety of staple price series for a variety of countries, on a monthly basis. All the prices used in the retrospective section of this work were downloaded using the GIEWS data and analysis tool.[6] It should be noted, however, that the commodities for which prices are available vary by country, as does the geographic unit of observation and the available date range. The starting date of the data is 2005 for most countries, but in some cases it is 2007. Where several prices were available for a given commodity in the same country (different locations or varieties), we compiled national averages. Where both retail and wholesale prices were available, we used retail prices because the majority of our samples are net buyer households. Finally, all prices were expressed in real terms and in local currency units.

The AGLINK-COSIMO *Agricultural Outlook* uses a global partial equilibrium model to predict yearly values of agricultural commodity prices in all regions of the world, using a ten-year moving window. At the time of writing, the most up-to-date available predictions reached 2018, which thus became our outlook year. The AGLINK-COSIMO does not model all countries of the world individually, so we had to use regions in some cases. Nicaragua, Panama and Guatemala, for example, were all considered to be part of the AGLINK-COSIMO region "CLA – Central Latin America".[7]

The household data from RIGA and the two price datasets were merged at the crop level. Since the availability of and definitions of crop data varied between datasets, we created categories of crops (modelled on the AGLINK-COSIMO commodity categories). Those categories are detailed in Table 2.2. Note that only staple crops were included in the study. One reason is that only staple prices are present in the GIEWS dataset, but it is not the only reason. Non-staple foods such as meats or fresh produce are more likely to be quickly cut out of a diet if too expensive. Since we do not allow for changes in consumption patterns in the model, our result would be weaker if we considered non-staple foods. As we have no information on feed prices, adding livestock on the production side would also have been ill-advised. Staple consumption, on the other hand, is fairly inelastic, and we think our method yields fairly robust results in this case. Before presenting those results, we illustrate the method using hypothetical price increases.

Table 2.2. **Commodity categories and codes**

Code	Stands for	Comprised of
WT	Wheat	Wheat
RI	Rice	Rice
CG	Coarse grains	Maize, sorghum, millet, barley, other cereals
OS	Oil seeds	Flax, sesame, palm nut...
HP	High protein	Soybeans, dry beans, peas...
RT	Roots and tubers	Potatoes, sweet potatoes, cassava, taro...

2.3. Hypothetical price shocks and illustrative results

In this subsection, we will illustrate the method using a theoretical 10% shock to the price of staples, as modelled in Zezza (2008). This allows us to illustrate our method and identify the relevant questions one needs to ask when analyzing price shocks.

As explained earlier, if the price of a commodity changes, the sign and size of the immediate welfare shock for a household ultimately depend on this household's net benefit ratio for that commodity. As noted already, the majority of rural households in many developing countries are net buyers of staples. Figure 2.1 illustrates this general tendency: of the nine countries selected for the study, only Viet Nam displays a positive net benefit ratio for staples in the rural sector.

What would be the immediate welfare shocks in the rural sectors of those countries if staple prices were to rise, for example, by 10% in each country? Figure 2.1 gives an answer at a glance, without even making calculations. Indeed, if we were to model a 10% price shock and calculate the welfare effect using equation (2.1), we would simply obtain values equal to a tenth of the NBR read off the bubbles in Figure 2.2. Thus, leftmost Bangladesh would display an estimated welfare shock of about –1%, and rightmost Viet Nam a positive shock of about 2.5%.

Figure 2.1. **Staple net benefit ratio (NBR) of the rural sector**

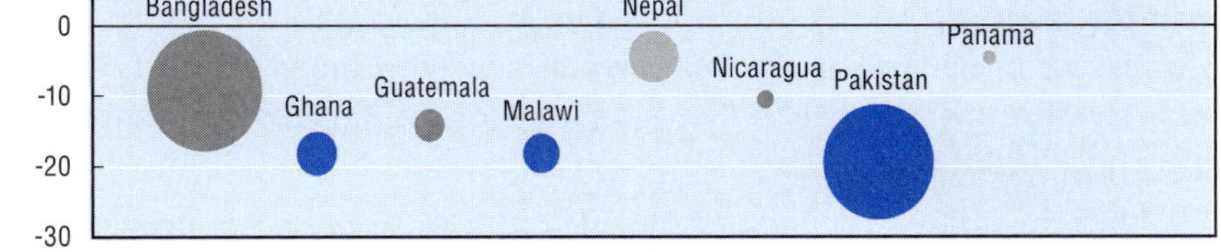

Note: Bubble size is proportional to the size of the rural population of the countries. Colours are light blue for values lower than –15%, dark grey between –15% and –5%, light grey if negative but above –5%, and dark blue if positive.

Source: Compiled from RIGA datasets.

StatLink http://dx.doi.org/10.1787/888932552475

Such figures might seem rather small, but recall that we are using a 10% increase for illustrative purposes only, the actual price shocks which hit

Figure 2.2. **Net benefit ratio (NBR) of staples by quintile of total expenditures (1 = poorest, 5 = wealthiest)**

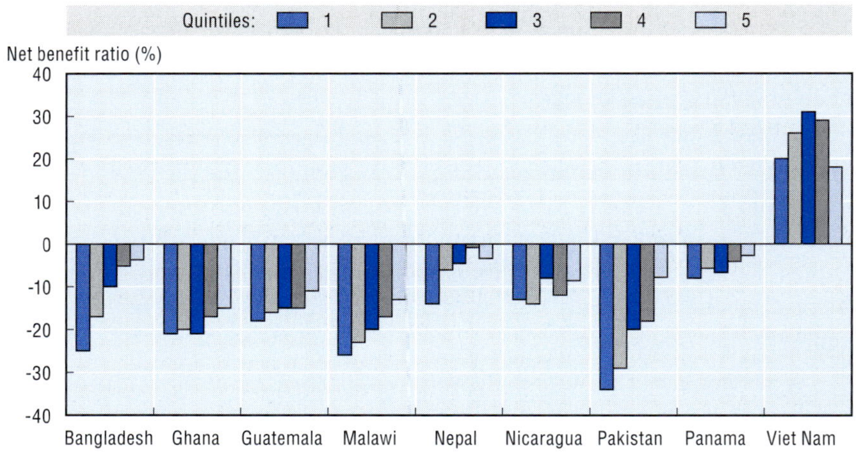

Source: RIGA datasets.

StatLink http://dx.doi.org/10.1787/888932552494

during the 2007/08 crisis were of much greater magnitude. The advantage of a single, across-the-board 10% hypothetical increase is that it can help identify vulnerable groups which, all else being equal, stand to be hit hardest by the food crisis, as in Zezza *et al.* (2008). Findings similar to theirs are presented in Figure 2.2, which again shows the NBR for staples, only now with households disaggregated by expenditure quintile. The chart reveals that the poorest quintile almost invariably displays the lowest NBR in their country (in two exceptions, it ranks second-lowest by a very small margin). This means that the poorest stand to be hurt most by – or in the case of Viet Nam, to benefit least from – a rise in staple prices. Again, welfare shocks from a hypothetical 10% increase will simply equal a tenth of the values read off the bars. Thus, for example, as a group, the 20% poorest rural Bangladeshi households (the leftmost bar of the graph) stand to perceive a welfare shock of about –2.3%, a value significantly larger than that estimated for the Bangladeshi rural sector as a whole (about –1%).

With the above-described figures we can already draw two conclusions: first, that the rural populations of developing countries are predominantly net staple buyers who will be hurt by high staple prices; second, that among them, the poorest will be hurt most. The work by Zezza *et al.* (2008) starts from the same observation, then pushes the analysis by documenting in depth the characteristics that identify the "poorest" households. Our work takes the analysis along different lines: modelling actual, rather than hypothetical, price shocks, and differentiating among various staples. Indeed, once we aggregated all staple foods together and found that all quintiles were net buyers, it should come as no surprise that the poorest quintile spends the most (in proportion) on staples. It may be that the results of Figure 2.2 do little more than documenting Engel's law.[8] As soon as one digs deeper, the results become much less homogenous and conclusions more equivocal than in this cursory analysis.

Further motivation comes from the fact that in Figure 2.2, the sign of the marketed surplus for staples is identical for all quintiles in each country: positive for all Vietnamese and negative for everybody else. This is surprising because one could think that richer farmers would be net sellers of food, whose surplus feeds the city dwellers and the poorer, net buyer segments of the rural population. Yet this does not appear to be the case in many developing countries. Another study of 15 RIGA datasets showed that the percentage of production that is marketed not only is low (ranging from 30% in Nigeria to 61% in Viet Nam), it also remains low in the largest landholding quintile (Brooks, Dyer *et al.*, 2008). Similarly, Figure 2.3 reveals that net sellers are a minority in all quintiles of all countries (except Viet Nam, where they are the majority in all quintiles). An explanation could be that the "net seller/net buyer" division does not exist at the national level but rather at the global level: the entire rural sector of countries like Viet Nam produces staple surpluses which feed the rural sector of net buyer countries like Bangladesh.[9] These considerations beg the question: what roles do different crops play in the overall result?

Figure 2.3. **Share of net sellers of staples in each expenditure quintile**

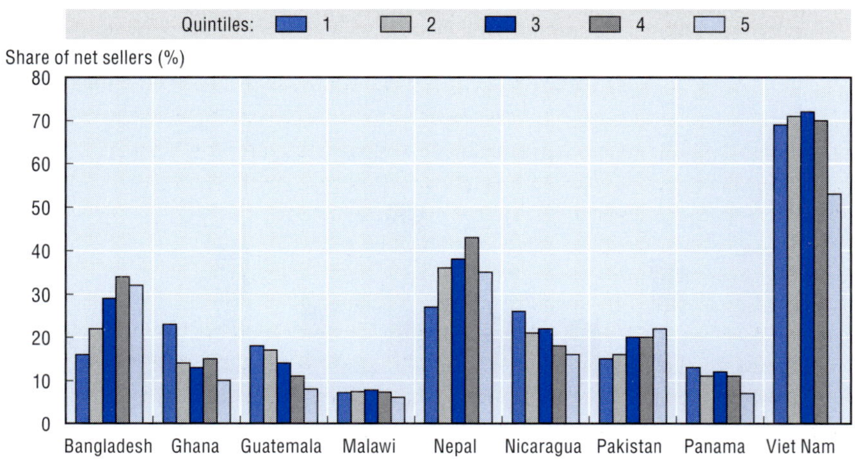

Note: Rural households only, weighted by population (1 = poorest, 5 = wealthiest).

Source: RIGA datasets.

StatLink http://dx.doi.org/10.1787/888932552513

This question leads to the core of the analysis: the crop-level impacts of welfare shocks over time. The methodology in what follows is similar to what has been presented so far, but we now focus separately on each type of commodity considered. Net benefit ratios (NBR) are obtained from the values of consumption and income related to each type of staple, both of which we present in Figure 2.4.

Figure 2.4. **Values of staple consumption (top) and production (bottom) as shares of total expenditures, by type of staple, in the rural sector**

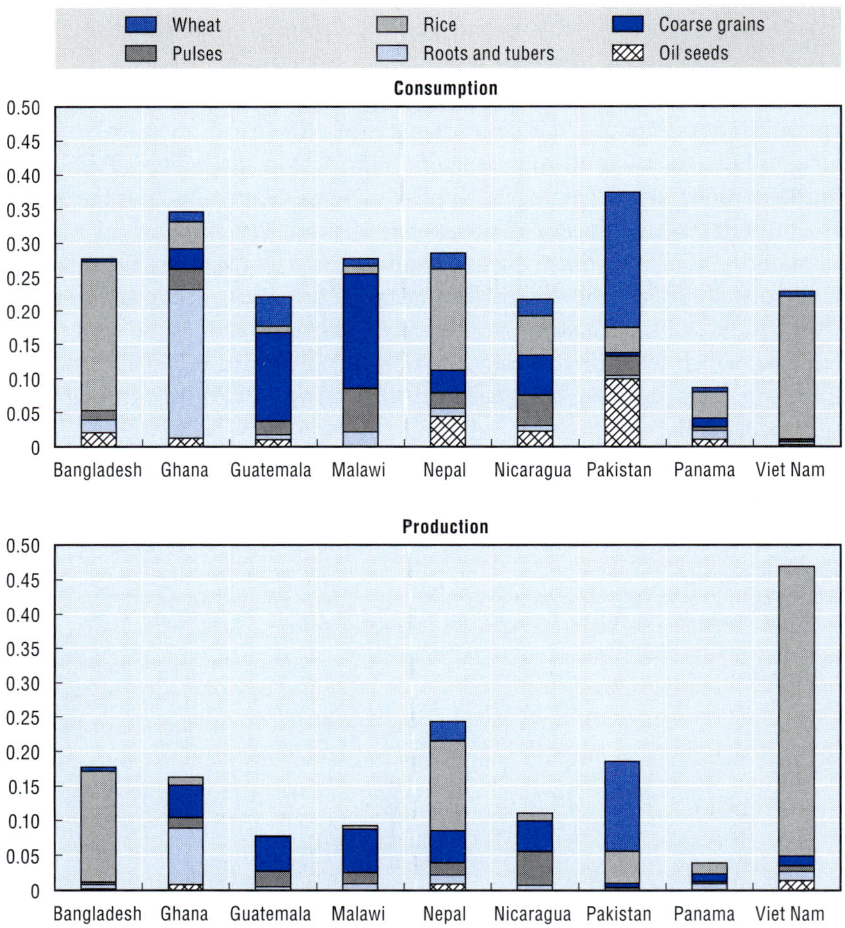

Source: RIGA datasets.

StatLink http://dx.doi.org/10.1787/888932552532

A few noteworthy observations emerge from the two pictures. First, there is great variation in the importance of staples as a whole (the size of the bars): in Panama, they barely account for 10% of expenditures; in Viet Nam, income from staples represents over 45% of total value of expenditures; in Pakistan consumption of staples accounts for 37% of expenditures. Secondly, the dominant staple varies by country (*e.g.* coarse grains in Malawi, rice in Nepal). Third, the degree of diversification is very different across countries. Bangladesh and Viet Nam almost exclusively produce and consume rice. In contrast, Nicaragua shows no clear dominant staple. This last observation in particular seems like it might matter when modelling actual price shocks by commodity. Indeed, a country with a diversified staple basis would be somewhat protected from severe welfare shocks as long as prices of all staples do not soar at once.

2.4. Retrospective look at the 2007/08 food price crisis

In this section we confront the LSMS data compiled with the actual price shocks that occured during the 2007/08 crisis. We first describe and discuss the price data, then simulate the shocks both at the peak of the crisis and as time-series.

GIEWS historical price data shows great diversity

The GIEWS database does not provide prices for each country and commodity that we wish to model. Table 2.3 presents the available price data. Despite this limitation, comparing Table 2.3 to Figure 2.4 shows that all major crops are available in the countries where they would seem to matter, with the exception of oilseeds (OS) which would matter somewhat in Pakistan, and pulses (HP) in Malawi. In Viet Nam, rice is so clearly dominant, that the lack of other price data is of less concern.

Table 2.3. **Prices available from the *GIEWS Database* for the countries of interest**

Country	Available price series
Bangladesh	Rice (RI), Wheat (WT)
Ghana	Rice (RI), Roots and Tubers (RT), Coarse Grains (CG)
Guatemala	Rice (RI), Wheat (WT), Coarse Grains (CG), Pulses (HP)
Malawi	Rice (RI), Roots and Tubers (RT), Coarse Grains (CG)
Nepal	Rice (RI), Wheat (WT)
Nicaragua	Rice (RI), Wheat (WT), Coarse Grains (CG), Pulses (HP)
Pakistan	Rice (RI), Wheat (WT)
Panama	Rice (RI), Wheat (WT), Coarse Grains (CG), Pulses (HP)
Viet Nam	Rice (RI)

Presenting the GIEWS price data for each country and each crop would be overwhelming. Instead, in Figure 2.5 we describe the evolution of staple prices in Guatemala, which provides an instructive example. Rather than a single

Figure 2.5. **Staple prices in Guatemala – 2005-09**

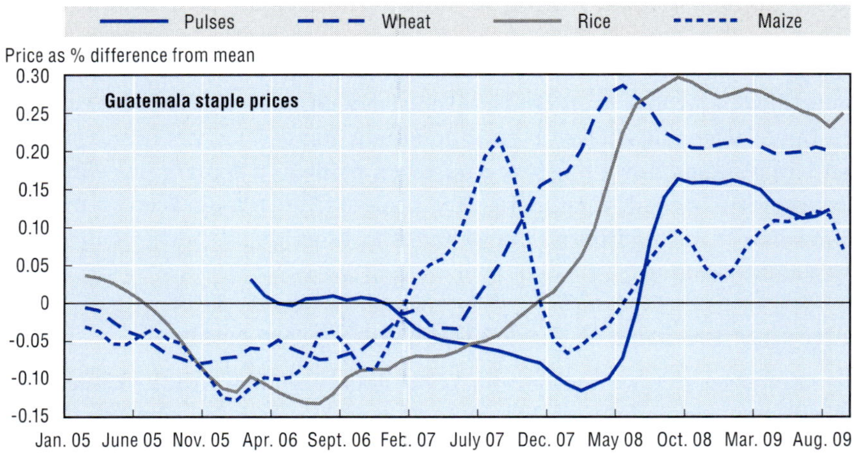

Source: Compiled from *GIEWS Database.*

StatLink http://dx.doi.org/10.1787/888932552551

"food price crisis" during which all prices would rise simultaneously then fall together, we observe that prices rose at different times for different crops over a two-year period. During the first observed shock (maize prices in January 2007), prices of pulses were actually falling and kept doing so until March of 2008. At that date, wheat and rice prices were rising sharply, but maize prices had fallen back to below-average levels. After July of 2008, prices of all staples had risen above average levels, and had hardly fallen back down a year later. This first glance suggests a picture which contrasts sharply with the one for world prices. While world prices all rose sharply in 2007 and decreased after mid-2008, the Guatemalan economy seems to be undergoing a more durable increase in the level of staple prices.

What are the likely effects of what we see for household welfare? As far as immediate shocks go, for a household that is a net buyer of all of these crops, it is not obvious at what point the decrease in the price of pulses will stop compensating for the increase in the price of other staples. An impact may be felt at any point during 2007 or 2008, depending on the relative importance of the crops in the household's budget. Once it hits, however, the negative welfare shock will be quite prolonged if nothing changes. By mid-2009 (supposedly after the crisis, as far as world prices are concerned), the four crops are all still 10% to 25% above their average level simultaneously. Our simulations help illustrate how those high prices can translate into welfare pressures.

The diversity of potential outcomes is demonstrated by a quick cross-country comparison. Since the only good for which data is available in all countries is rice, we use it as an example-crop and plot the evolution of the price of rice in all nine countries over the period 2007-09 (Figure 2.6). The lack of a discernible pattern across the panel is telling: while all countries saw a peak in the price of rice at some point in 2007/08, no two countries display the same evolution. Some countries saw an abrupt spike (Pakistan), others a slower increase (Nicaragua); some spikes were short (Viet Nam), others prolonged (Bangladesh); some early (Panama), others late (Malawi). Bangladesh saw a sustained increased price for the entire 2008 year, but the price returned to normal levels in 2009. In contrast, the price of rice in Ghana increased steadily and the available data does not yet show it falling back. In Viet Nam, the spike was both sharpest and shortest of all countries. Such differences are bound to induce dramatically different consequences for the various countries. Similar graphs (not presented here) were compiled for the other groups of staples, and similar diversity was observed across countries.

The analysis of those price evolutions leads to the question of how transient this "price crisis" really was, since most countries in our sample did not seem to recover. This question is worth keeping in mind, because the consequences of a structural break in commodity markets are quite different from those of a one-off crisis. If the new market conditions are to last, rural economies may need to adjust, and it might be that they will face continued depressed welfare levels until they do so. The policy responses in that case are also very different from the ones necessary to curb a transient crisis.

This quick overview of price evolutions in various countries for various crops only reinforces our conjecture made at the end of Section 2.2: that if we look at differentiated crops rather than staples in general, at actual price shocks rather than hypothetical ones, and at time-series of prices rather than point data, our cursory analysis of welfare shocks may turn out to be greatly insufficient. The diversity of welfare impacts across countries and household groups is likely to be substantial, not only in terms of magnitude but also in terms of timing and duration. We now explore both of these aspects.

Magnitudes, timing and duration of welfare shocks varied substantially across countries

Magnitude of welfare shocks at the peak of the crisis

The word "crisis" connotes a shock of particular brutality: sudden, unforeseen, and of large magnitude. Yet the evolution of rice prices shown in Figure 2.6 did not necessarily fit this description. There is no clear criterion by which to decide whether a peak in prices qualifies as a "crisis". In order to compare the severity of the "price crisis" across countries, Dawe and Morales-Opazo proposed to look for the point of "largest shock", identified as the greatest year-to-year increase in the price of any given staple (Dawe and Morales-Opazo, 2009). We used a similar method to identify the date of largest shock,

Figure 2.6. **Evolution of the price of rice in selected countries, expressed as per cent increase or decrease relative to the mean**

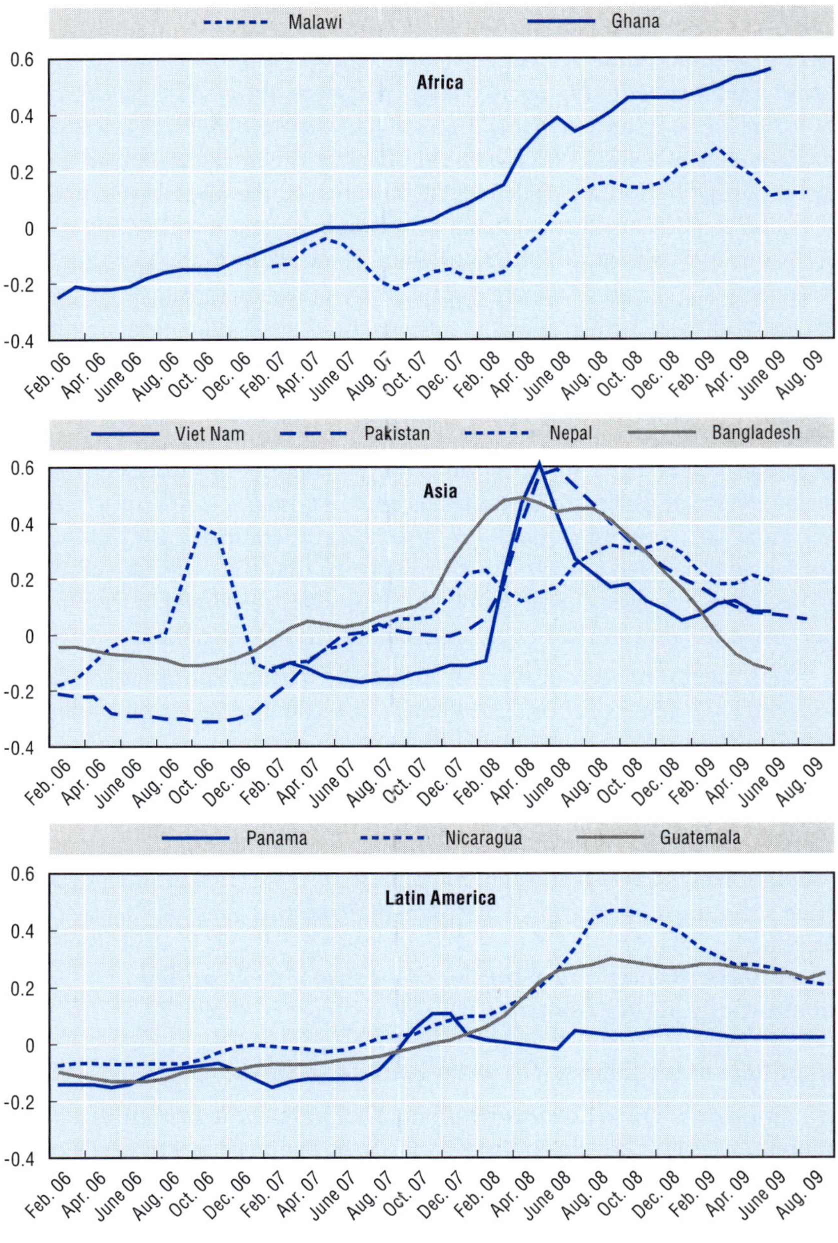

Source: Compiled from *GIEWS Database*.

StatLink http://dx.doi.org/10.1787/888932552570

summing price differentials across all staples using a six-month basis.[10] This "crop-cumulative six-month price differential" can be thought of as the worst of the crisis, when staple prices, cumulatively, rose sharpest over a 6-month period. The timelines of cumulative shocks are plotted in Figure 2.7, which shows that dates of sharpest increases are clearly identifiable for some countries but less obvious for others. Those dates and the corresponding shocks are provided in Table 2.4.

Table 2.4. **Dates of the largest cumulated price shocks by country (shocks over six-month period), and details by crop, %**

	Date of largest shock (all crops combined)	Size of shock by crop				
		Rice	Wheat	Coarse grains	High protein	Roots and tubers
Bangladesh	Feb. 08	34	32			
Ghana	Jul. 08	24		80		42
Guatemala	Aug. 08	21	2	15	29	
Malawi	Mar. 08	8		110		-9
Nepal	Oct. 06	55	18			
Nicaragua	Oct. 07	5	.	30	88	
Pakistan	Jun. 07	43	1			
Panama	Oct. 08	3	8	32	24	
Viet Nam	May. 08	85				

Source: Compiled from *GIEWS Database.*

At the point of the largest price shock, how hard was the blow in terms of welfare? Figure 2.8 presents some of the key findings of our analysis: they are the result of applying the real price shocks from GIEWS to the household data from RIGA. We present them as a matrix of graphs, where each country graph represents the welfare shock by expenditure quintile, at the date of the sharpest six-month price shock. Households are disaggregated by expenditure quintiles, and crops are disaggregated as much as data permitted. For each country and each quintile, the fully coloured dots represent the impact due to a particular staple, while the hollow circle represents the overall welfare impact. Analyzing the nine graphs gives us insights into how each country experienced the food price crisis.

First, consider the hollow circles and the magnitudes of shocks. As expected, in some cases the welfare impact is much larger than that of the hypothetical 10% price increase simulation. Some household groups in some countries reveal shocks that are rather severe, such as the poorest quintile in Malawi (25% drop in welfare in six months), or some groups in Bangladesh, Nepal, and Ghana that show a drop of about 7%. On the other hand, Panama, where the most dramatic shock was only –0.6%, stays virtually unaffected. Viet Nam stands out as the only country where all rural households gain.

Figure 2.7. **Crop-cumulative six-month price differentials**

Per cent increases over value six-months earlier, summed across all available crop prices

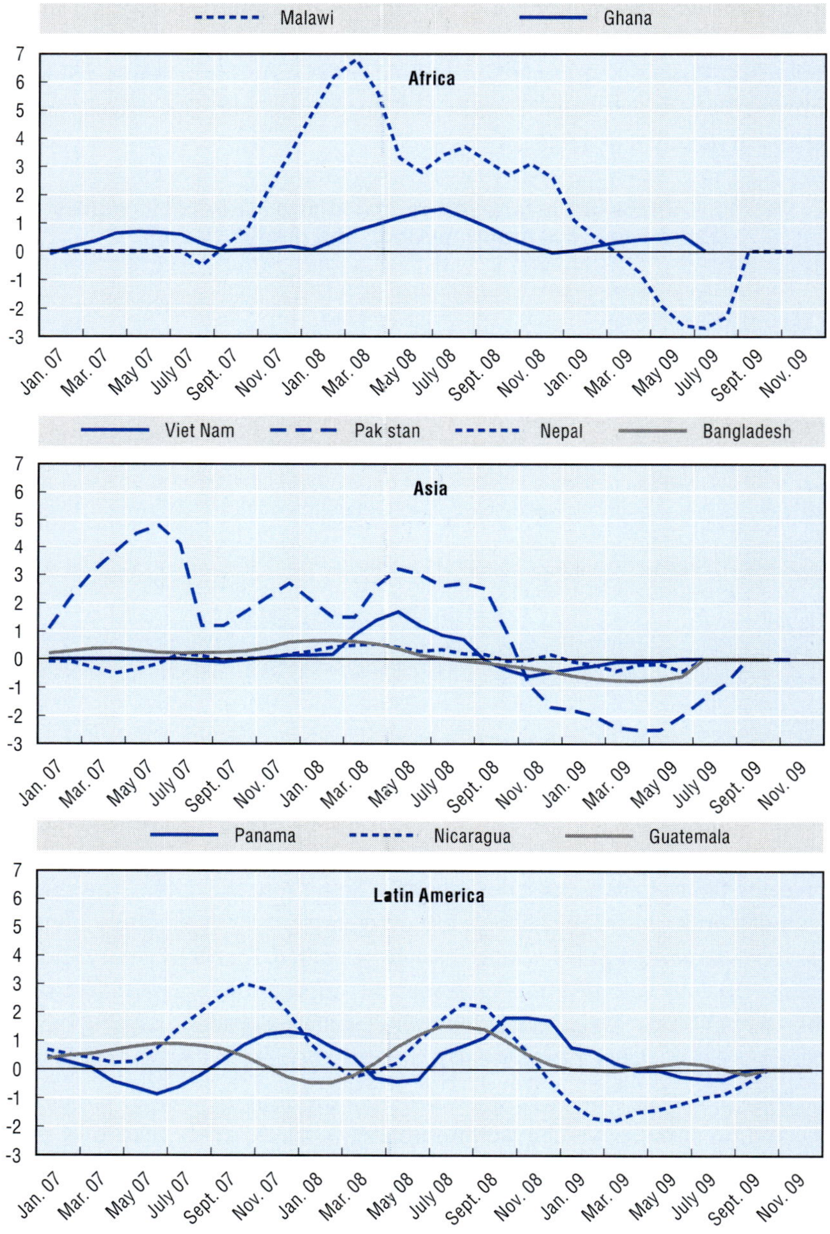

Source: Compiled from *GIEWS Database.*

StatLink http://dx.doi.org/10.1787/888932552589

Figure 2.8. **Welfare impacts by crop, at the time of the "worst" price shock (largest crop-cumulative six-month price increase)**

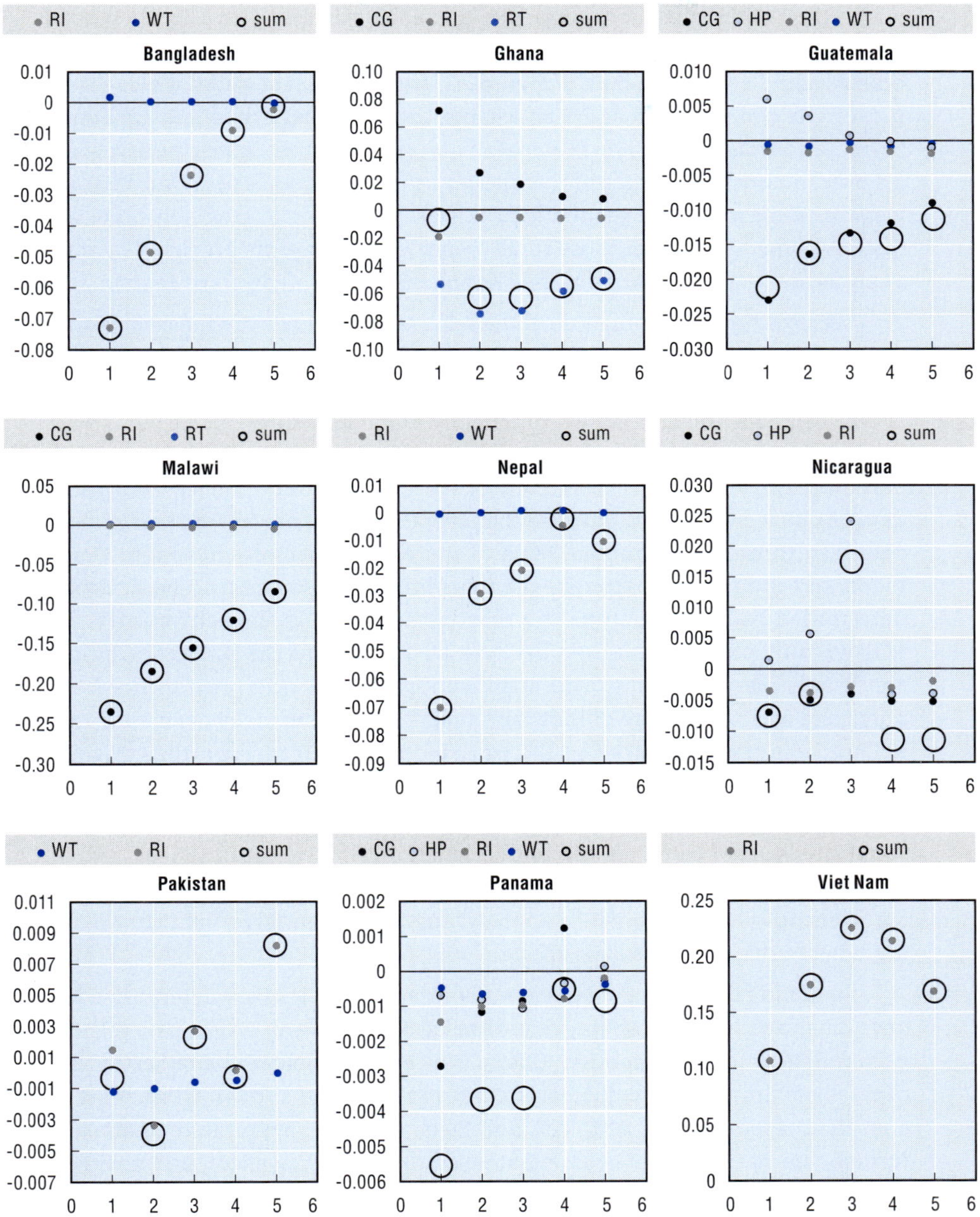

StatLink http://dx.doi.org/10.1787/888932552608

The disparities between quintiles are also worth observing. The hollow circles in countries like Bangladesh or Malawi show a clear linear pattern by which the poorest are hit the hardest, and each quintile fares slightly better (less worse) than the previous one. That pattern is virtually identical to the one we could see on the bars of Figure 2.2. Nepal's poorest quintile also loses most in its country, with a –7% shock, but the linear pattern is less apparent there. In contrast, the poorest Ghanaian quintile fares best in its country, with an overall shock of about zero, while all other quintiles lose around 5% in welfare. Meanwhile, the results for Pakistan and Nicaragua exhibit no clear distributional pattern. For those three countries, the impacts predicted by this analysis are very different from the ones of our cursory analysis of Section 2.2.

We now turn to the analysis of the relative importance of each crop in the welfare impacts (the coloured dots). First, we identify a group of countries for which only one crop seems to have mattered, easily identifiable because the "overall impact" hollow circles practically encircle a single dot each. From Figure 2.4 we already knew that in Bangladesh and Viet Nam rice would be the only crop that really matters (and in Viet Nam, it is the only crop we have price data on), so it comes as no surprise that overall impacts are fully driven by rice values in those two countries. Malawi and Nepal are more diverse in their foods and production patterns, yet their impacts also seem to be entirely determined by a single crop. The reasons are specific to each country. In Malawi, only one crop matters because at the time of the maximum price shock only one commodity was significantly affected: the price of rice had only gone up 7%, the price of roots and tubers (RT) had actually decreased 8%, while coarse grains had seen a 110% price increase in six months, thus driving the welfare impact. In Nepal, the situation is completely different: rice and wheat both saw relatively large increases in price, and both represent a significant part of consumption and production. However, it turns out (graph not shown) that all five quintiles of the rural population are virtually self-sufficient in wheat, with NBR's approaching zero: even large fluctuations of the wheat price will have little effect.

Switching the focus to countries where welfare shocks are determined by several crops, we see that price shocks on different crops can either accumulate or offset each other. Ghana is a clear example of the latter. At the date of maximum shock, prices of coarse grains, rice and RT all increased sharply (respectively by 80%, 42% and 24%). It also happens that Ghanaian farmers are net sellers of coarse grains, so the 80% price shock had positive welfare effects, and helped to partly offset the negative effects of the rice and roots and tubers price shocks. In the case of the poorest quintile, the net effect is close to zero, making them the least ill-affected group. Nicaragua also offers an interesting case. While we had seen that all rural quintiles are net buyers of "staples" in general, this is not the case for pulses in particular. The three poorest quintiles are net sellers of pulses, and the two wealthiest ones are net buyers. Therefore, when prices on pulses increased 88% over six months, this was good for the three poorest quintiles, helping them offset the negative impact of the simultaneous 30% increase in CG prices. Meanwhile, it made things worse for the two wealthiest quintiles. For the middle-income rural households of the third quintile, this even leads to an unexpected sign reversal, as they perceive a positive overall welfare effect of almost 2%.

This subsection leads us to a few conclusions about the magnitudes of welfare shocks. First, we found that even when focusing on each country's most severe price shock over the past four years, the magnitude of the actual impacts on welfare shows large disparities between countries. Second, we found the interplay of different price shocks, and differences in the composition of crops in households' production and consumption, significantly complicates the picture. The cursory analysis of Section 2 gave incorrect predictions in three cases out of nine.

Analyzing welfare over time

The previous sections showed that the relative importance of crops in the portfolios of households, and the relative size of the price shocks on crops, played a defining role in the welfare impacts induced by the price crisis. With good understanding of the forces involved, we can now add timing into the equation and analyse welfare timelines in all nine countries.

Figure 2.9 shows, for the five expenditure quintiles of each country, the evolution of the welfare incidence during the past few years (in percentage difference from the mean level), which we call the "welfare path". Note that we used as long a time frame as was available for each country, such that some charts span four years, others three, and others only two. Again, the first impression that emerges from the chart matrix is one of diversity. Some countries show minimal disturbances (Panama), others sizeable deviations from the mean (Malawi, Viet Nam). All countries show idiosyncratic welfare paths.

There are important between-quintiles disparities in terms of welfare impacts. Looking at Guatemala or Nicaragua, we see that all the welfare paths of households are tightly packed together, suggesting that rich and poor all go through equally good or bad times as agricultural prices fluctuate. That is not the case for Malawi and Bangladesh. In those countries, the negative shock was accompanied by a clear widening of welfare disparities. Interestingly, Panama shows the opposite phenomenon (though to a lesser extent): there, the hard times were associated with narrower gaps between quintiles. Looking further at the differences between shocks on various quintiles, we find four countries for which the negative shock was hardest on the poorest quintile (*i.e.* the full line is the lowest during the negative shock). In three

Figure 2.9. **Evolution of welfare over time, in percentage difference from the mean**

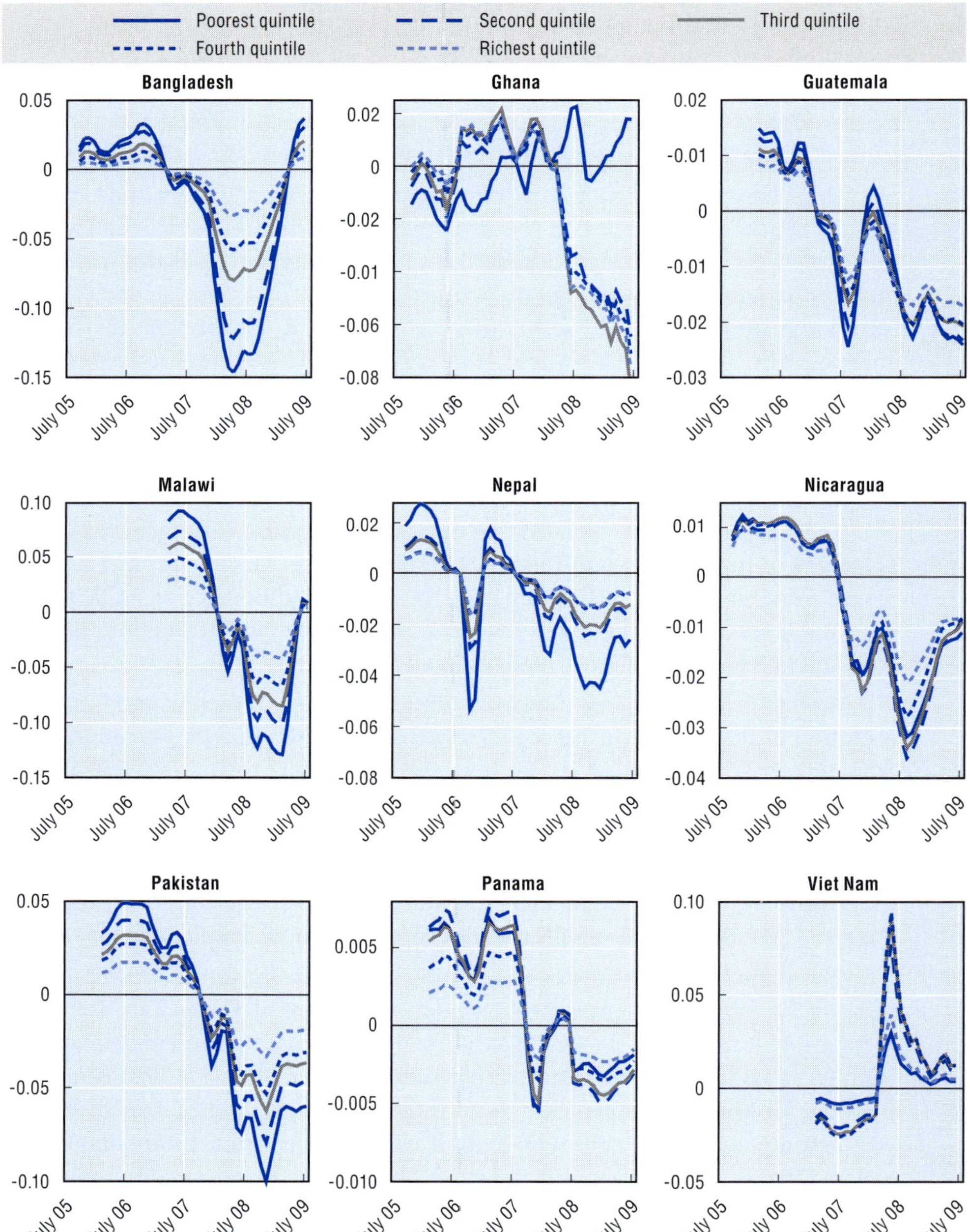

StatLink http://dx.doi.org/10.1787/888932552627

countries the poorest fared no worse than others; in one case, they fared best in their country. According to these observations, the food crisis cannot be said to have widened disparities, and it cannot be said to have been hardest on the poor segments of the rural sector. These findings contrast with the predictions of the cursory analysis in Section 2.2, which led us to think that almost without exception the poor would be the biggest losers when prices rise. As we mentioned earlier, this result may simply have been an illustration of Engel's law and was likely *too* cursory.

Let us finally make a few remarks about the timing and duration of the "welfare crisis" (rather than just the food crisis). The onset of the sharpest welfare impacts varies from late 2006 (Nepal) to mid-2008 (Ghana). This echoes our earlier observations concerning the timing of price shocks. Comparing the duration of the crisis between countries is a difficult task. Bangladesh is the only country that undeniably shows a transient "welfare crisis": welfare drops dramatically and suddenly, and then rises again to average levels. Viet Nam shows an equally clear temporary "welfare boom". But recall that those two countries were precisely the two that virtually produce and consume only a single staple (rice). As soon as the production and consumption patterns become more complex, so does the welfare path. Malawi and Nicaragua also display rather identifiable "crises" in 2008/09. Nepal shows a sharp and transient drop in welfare in 2006, a return to normal, then a less obvious "crisis" at the end of 2008. Pakistan seems to be going through a gradual deterioration of welfare, which stays below average during 2009. Panama and Guatemala both show no signs of recovery after the shock, and four out of five Ghanaian quintiles fare worse than ever at the end of the time-series. The welfare paths we observe clearly suggest that in most countries the effects of the rising food crops were prolonged and were not necessarily reversed at the "end" of the crisis.

We thus conclude our retrospective part of this chapter with the observation that the "past" crisis was not necessarily over even by the middle of 2009. This indicates that we may be witnessing some deeper structural change rather than a temporary crisis. Moreover, as the next section points out, prices are largely predicted to continue rising over the next ten years.

2.5. Prospective look at the 2009-18 period

In this section, we switch to a forward-looking analysis, using the AGLINK-COSIMO price predictions to plot out the welfare effects of future price trends. The AGLINK-COSIMO price data has the advantage of being very complete, such that we are less limited by data availability than we were with GIEWS. We now have price predictions for all crops of interest and all countries, with the exception of pulses, which are not modelled in AGLINK-COSIMO. This may bias somewhat our results in Nicaragua, but likely will not matter dramatically for most other countries of the sample.

The price predictions of AGLINK-COSIMO over the next few years show as much complexity and variation as the GIEWS price data did, and there is no need to discuss them in detail. Each country shows unique price evolutions which do not simply follow the international prices mentioned in the introduction. In the long-run, however, a clear pattern emerges: in 2018, prices of all commodities in all countries are predicted to have risen substantially (by about 20 to 80% in general, but up to 130% for maize in Malawi).

We simulate the effect of those prices on welfare. Because we are using medium-term data, the values of our "immediate welfare effects" should be interpreted with caution. These results should not be seen as portraying actual welfare outcomes, but rather pressures that commodity markets will exert on welfare – which can provide insights in terms of the direction of likely adjustments. In the long term, for example, some farm households that lose immediately from higher prices may be able to benefit by transforming themselves into net sellers.

We chose to disaggregate welfare paths by crop (rather than by household quintiles like earlier), so as to observe which crops will have the strongest impact on household welfare and household behaviour (although we cannot model the latter). Our simulation results are presented in Figure 2.10. The graphs feature the paths of welfare pressures due to each crop, and the overall welfare effect (thick line). All are calculated using predicted price differentials from their initial 2009 level.

We observe three general shapes of graphs. The first, observed in Malawi, Pakistan and Guatemala, shows straight and uninterrupted downward pressure on welfare over the whole period. The second, observed in most other countries, features a more-or-less pronounced welfare increase in the first year, followed by an uninterrupted drop. In those cases, the welfare path drops below the original 2009 level in 2012 at the latest. The outlying shape is that for Viet Nam, which shows a mirror image of what we just described: a two-year drop followed by a continuous increase.

Countries which improve from 2009 to 2010 might be the ones that, in our retrospective analysis, had ended on a low level in 2009. Could it be that these countries will finally pick up from 2010? That is the case for Ghana and to some extent in Panama. In Guatemala and Pakistan, on the other hand, welfare keeps falling as it did in the past. These results, again, point to the rise in prices as a lasting phenomenon. They also confirm our fear that some rural sectors which have not recovered from the so-called 2007/08 price crisis, are not likely to do so unless livelihood strategies adapt.

Figure 2.10. **Forward-looking welfare effects – in differences from 2009 level**

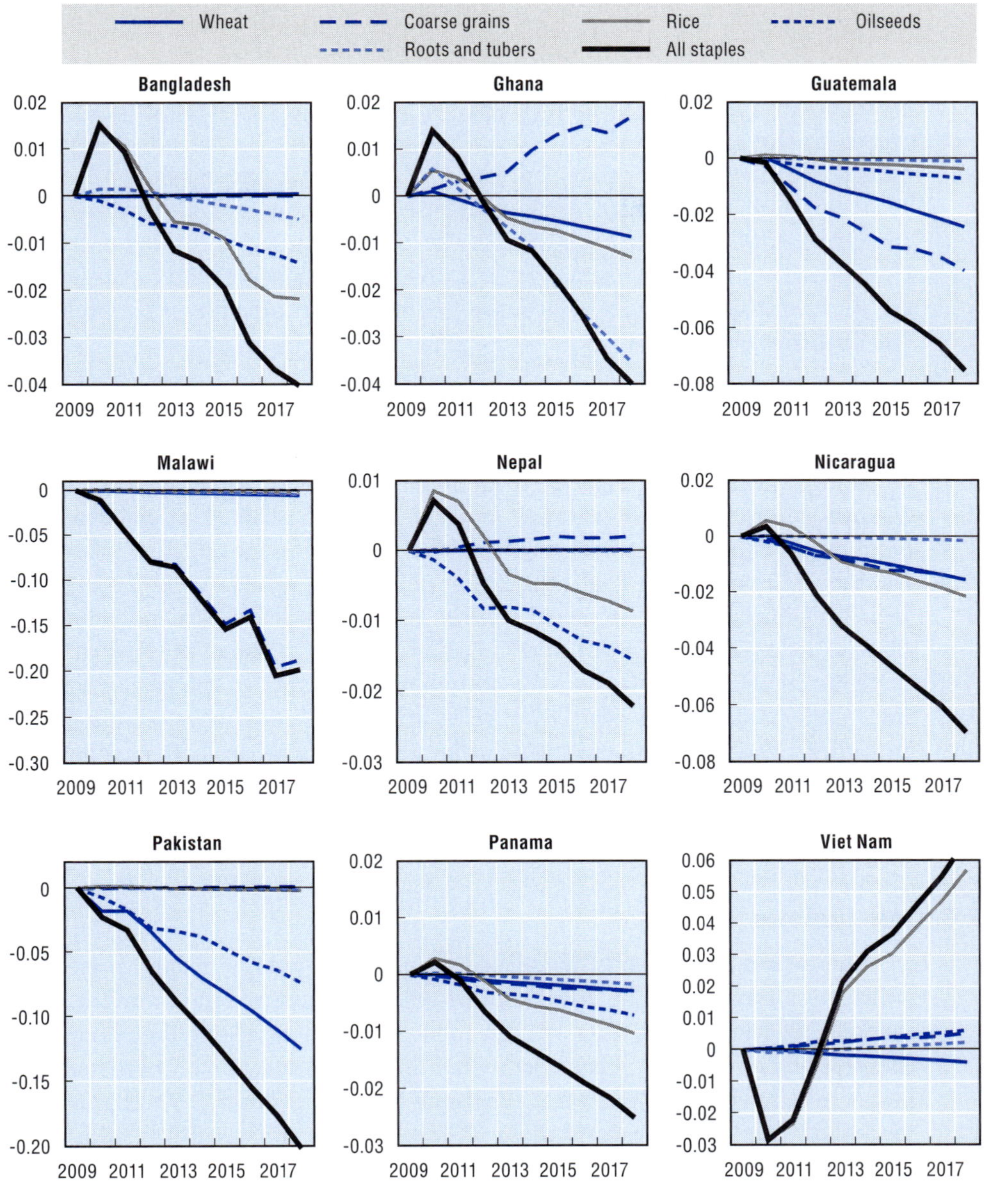

Note: The black, thick line represents the overall welfare path, while thinner lines represent one staple each.

StatLink http://dx.doi.org/10.1787/888932552646

The magnitude of the welfare pressures covers a wide range. In Pakistan and Malawi, welfare falls by 25% and 20%, respectively, ranking them among the biggest losers again. The variations in Bangladesh, however, do not exceed 6% in magnitude, whereas they were hit the hardest of our sample in 2008. In contrast, Panama had seen no variation larger than 0.7% which means they had been relatively unaffected, but by 2018 the downward pressure on welfare reaches –2%.

The contributions of each staple to the welfare shock are similar to those we made using GIEWS data. Bangladesh, Viet Nam and Malawi still seem to be driven by the price of a single crop, while other countries are affected by cumulative effects. In Nepal, the average welfare effect is driven by the prices of rice and oilseeds, which work in opposite directions at first, but then coincide in lowering the overall welfare effect. Interestingly, at the end of the forecast period, in 2018, almost every crop in every net buyer country works to the disadvantage of the overall welfare (coarse grains being an exception in two countries only).

In conclusion, unless households and markets adjust, the staple price predictions seem rather unfavourable to rural households of developing countries. In the long run, we predict downward pressure on welfare for eight out of nine countries. Yet there is no doubt that households will try to adapt to the new commodity markets. They may adapt their consumption habits or production volumes in response to the shifting incentives; the question is how fast they will do so. One might think that households would quickly change their consumption habits to avoid a crisis. However, since we are modelling staples (the cheapest of foods), and since all staple prices rise in the long run, there is actually very little room for households to adapt. On the production side, while adaptation is typically slow in agriculture, nine years should be enough for households to substitute away from unproductive activities and invest in crop production to benefit from the high prices. This may be a more likely scenario, and a more optimistic one, as it could turn the original negative welfare shock of 2008 into sustained growth opportunities.

2.6. Conclusions

The world has recently witnessed dramatic variations in food prices which may have been the early sign of structural change in commodity markets. Because the effects of high food prices on rural livelihoods are complex and ambiguous, it is unclear what the welfare impacts of these high prices may have been, and what they may be in the future. This chapter proposed a method to evaluate those welfare impacts, and used it in both a retrospective and a prospective analysis of welfare impacts in nine developing countries.

Our approach bridges two kinds of economic studies: those that focus on markets and prices, and those that focus on household data. For example, a study like that of Dawe and Morales-Opazo (2009) documented the extent of the price shocks during the crisis, but could only hypothesise the consequences by stating that "such large price increases almost certainly had severe impacts on the effective purchasing power of the poor". In contrast, a study like that of Zezza *et al.* (2008) simulates a 10% staple price increase, and effectively condenses down to a comparison of staple net benefit ratios along various household characteristics. By bridging the two approaches, the results should give a better indication of the likely impacts.

The immediate impression is one of diversity. This diversity can clear up some misconceptions about the welfare implications of food price crises. Section 2.2 set out with a cursory analysis which produced very general results, similar to those from previous studies, but which failed to hold when detail was added to the picture.

We initially found that rural households in all but one of our countries of interest are net buyers of staples in general, but later showed that they are not net buyers of *all* staples, and that the sign of their net balance for any given crop varies by expenditure quintile. Several households in several countries were shown to be net buyers of one staple and net sellers of another. Those households' welfare impact will depend on the relative size of the price increases between various staples, and cannot be predicted using a general "staple" category. Using actual price data at the crop level, we identified two household groups which fared dramatically better than we had originally predicted.

Another general result from our cursory analysis was that the households from the poorest quintile invariably fared worst in their country (or second worse in a single exception). However, once we had disaggregated the analysis by crop and introduced actual price shocks, this was not found to be a general result. We could not document a systematic increase in welfare disparities either, and they were even reduced in one case.

Those observations thus reinforce our belief that failing to focus on the crop-level may lead one to overlook non-trivial effects and to oversimplify conclusions. The relative timing of price shocks for various crops and the relative magnitude of those shocks both matter to the analysis and generate substantial diversity in the estimated welfare impacts.

Despite this diversity, we can draw a few general conclusions from both our retrospective and prospective analyses. First, the 2007/08 food price crisis carried well into 2009, and may still not be over in some countries. This is particularly worrisome, as households that suffered month after month of depressed welfare may have reached a dire situation, which could be difficult to recover from. Second, the conclusions from our forward-looking analysis point to continued increases in staple prices, the consequences of which will depend on whether rural households are able to turn these high prices to their advantage.

The short-run policy implication from the above observations is that pre-emptive action must be taken as early as possible to contain the impacts of another food crisis. Retrospective analysis tells us that many countries were still seeing low welfare outcomes as of 2009, and prospective analysis suggests that, in the absence of adjustment, another round of falling welfare is likely in the coming years. Those hit hardest by the first crisis may still be recovering from its long-lasting effects (for example, if they sold assets, took children out of school, developed nutritional deficiencies or sank into debt). A few household quintiles of our sample saw welfare drops of over 10% in the first crisis, for several months in a row: safety nets must be put into place before the next crisis strikes.

The rise in prices is predicted to be sustained in the long run. This could be seen as a long-term problem, in which case crisis relief is hardly a sufficient response. Protective policies with long-term impacts might also need to be put into place to help struggling households meet their food budget. On the other hand, high food prices also offer opportunities for the rural sector. Notice that Viet Nam, a success story of the green revolution, is the outlying country of our sample, the only one that benefits from high food prices. Investments could be made to facilitate adaptation of rural households to the new price conditions. Any household that manages to reach net supplier status will gain rather than lose from the future rise in prices and government policies can help induce this change.

We already hinted at a few caveats which could qualify our conclusions. One is the problem of data quality: it is often believed that households bias their survey answers to appear poorer than they are, in which case we may be overestimating the hardships induced by expensive staples. Another is the absence of second-order effects in our methodology. Higher staple prices could, for example, increase demand for hired labour on net seller farms, increase off-farm income, and thus mitigate the negative welfare shock. In contrast, higher prices on staples could also drive up the prices of other foods. The links between staple markets and feed markets are strong, and meat prices may rise substantially, thus exacerbating the negative welfare shock. Such second round effects are considered in the DEVPEM analysis of Chapter 3.

To summarise, the recent increase in staple prices created downward pressure on welfare in all but a few cases. The magnitude and timing of those welfare shocks depended on the type of crops produced and consumed by each rural household, and overlooking these details masks important nuances and substantial variation across countries. Price predictions indicate that the rise in prices may not have been a transient incident, but rather may be marking a structural break in commodity markets, with higher future prices threatening welfare but also creating opportunities for those who can increase staple production.

Notes

1. The authors express their gratitude to Alberto Zezza, Carlo Azzarri, Luca Tascotti, Ana Paula de la O Campos at FAO for their work with the RIGA data, as well as Jonathan Brooks, Andrzej Kwieciński, Pete Liapis and Michael Plummer at OECD for their insightful comments.
2. At the time of writing, more information on RIGA datasets could be obtained from *www.fao.org/es/ESA/riga/english/index_en.htm*.
3. At the time of writing, more information on GIEWS could be obtained from *www.fao.org/GIEWS/english/index.htm*.
4. At the time of writing, more information on AGLINK-COSIMO could be obtained from *www.oecd.org/document/6/0,3343,en_36774715_36775671_40969158_1_1_1_1,00.html*.
5. Zezza *et al.* report computing second-order welfare effects using own-price elasticities of supply and demand, and finding no qualitative differences for their analysis.
6. At the time of writing, GIEWS data was available at *www.fao.org/GIEWS/pricetool/*.
7. We used GIEWS data to check how valid this approximation was. Correlations between the evolutions of prices (as differences from their mean levels) in Guatemala, Nicaragua, and Panama were 0.42 on average. They were strong for wheat and rice prices. The outstanding exception was for pulse prices in Nicaragua, which were negatively correlated with the two other countries (around –0.4).
8. Engel's law originally states that the income elasticity of food (not staples) is less than one. However, staples being inferior goods, the observation is likely to be even more dramatic when focusing on staples alone. This might be the reason why a cursory analysis points so unequivocally at the poor being the biggest losers in a food price crisis.
9. Other explanations could involve data errors: If respondents chronically over-report their expenditures and under-report their incomes, they will tend to appear as net buyers. It could also be that we over-value home consumption and food gifts (valued at the market price for purchased goods) possibly inducing a sizeable error in our estimation of staple expenditures.
10. With our sample of countries and goods, year-to-year increases seemed to miss the most dramatic shocks; we thus decided to shorten the time frame.

References

Barrett, C.B. and P.A. Dorosh (1996), "Farmers' welfare and changing food prices: nonparametric evidence from rice in Madagascar", *American Journal of Agricultural Economics*, Vol. 78, pp. 656-669.

Brooks, J., G. Dyer *et al.* (2008), "Modelling agricultural trade and policy impacts in less developed countries", *OECD Food, Agriculture and Fisheries Working Papers*, No. 11, OECD, Paris.

Budd, J.W. (1993), "Changing food prices and rural welfare: A nonparametric examination of the Cote d'Ivoire", *Economic Development and Cultural Change*, Vol. 41, pp. 587-603.

Carletto, G., K. Covarrubias *et al.* (2007), "Rural Income Generating Activities (RIGA) study: Income aggregate methodology", Agricultural Sector in Economic Development Service, FAO, Rome.

Dawe, D. and C. Morales-Opazo (2009), "How much did developing country domestic staple food prices increase during the world food crisis? How much have they declined?", *ESA Working Paper*, No. 09-09, FAO, Rome.

Deaton, A. (1989), "Rice prices and income distribution in Thailand: a non-parametric analysis", *The Economic Journal*, Vol. 99 (Supplement: Conference Papers), pp. 1-37.

FAO (2009), *Global information and early warning system – on food and agriculture* [GIEWS], December 2009.

FAO (2009), *The state of food insecurity in the world*, FAO, Rome.

IFAD (2001), *Rural Poverty Report 2001*, IFAD (International Fund for Agricultural Development).

Ivanic, M. and W. Martin (2008), "Implications of higher global food prices for poverty in low-income countries", SSRN eLibrary.

OECD-FAO (2009), *OECD-FAO Agricultural Outlook 2009-2018*, OECD-FAO, Paris.

World Bank (2008), *The World Development Report 2008 – Agriculture for Development*, No. 366, The World Bank, Washington, DC.

Zezza, A., B. Davis *et al.* (2008), "The impact of rising food prices on the poor", unpublished manuscript, FAO, Rome.

Chapter 3

The Distributional Implications of Agricultural Policies in Developing Countries – Findings from the Development Policy Evaluation Model (DEVPEM)

by

Jonathan Brooks, Mateusz Filipski, Erik Jonasson and J. Edward Taylor[1]

This chapter presents the Development Policy Evaluation Model (DEVPEM), a new simulation model which captures four critical aspects of rural economies in developing countries: 1) the role of the household as both a producer and a consumer of food crops; 2) high transaction costs of participating in markets; 3) market linkages among heterogeneous rural producers and consumers; 4) the imperfect convertibility of land from one use to another. The results of simulations for six country models show that no untargeted agricultural policy intervention is pro-poor within the rural economy. While agricultural policy instruments are less efficient at raising rural incomes than direct payments, the degree of inefficiency of some market interventions, notably input subsidies, is not inevitably as high as observed in developed OECD countries.

3.1. Introduction

A major aim of government policies in developing countries is to raise the incomes of agricultural and rural households. In developed economies, the most effective way of addressing low incomes among farmers, or any other section of society, in the short- to medium-term is via social policies. Agricultural policies in general, and market interventions in particular, are shown to be relatively inefficient at transferring incomes to farmers (OECD, 2001). In the long term, agricultural incomes can be strengthened by measures to improve competitiveness, which can be done by supporting the enabling environment in general, and via an agricultural policy that focuses on the provision of public goods and the correction of market failures (OECD, 2002).

In the case of developing countries, plausible reasons have been advanced for why this prescription may not always be valid. In the first place, effective systems of social protection may not exist and – pending their development – there may be a need to look for short-term alternatives, including the use of agricultural policies. When international food prices spiked in 2007-08, a number of governments responded by mitigating the pass-through of price increases to the national level (Demeke *et al.*, 2009; Jones and Kwiecinski, 2010). Input subsidies have also been suggested as a way of achieving a rapid impact on production and incomes, while simultaneously providing a bridge to higher incomes by enabling farmers to break out of poverty traps (Dorward, 2009). Essentially, market interventions that offset the effects of market failures (for example in credit markets) are seen as a practical alternative to policies that correct them at source.

A key piece of information needed to inform this debate is how effective agricultural policies are at raising incomes in the short to medium term. The finding that they perform poorly in OECD countries is based on results from the OECD's Policy Evaluation Model (PEM), a partial equilibrium model in which output and factor markets are linked, and the effects of policies on farmers' incomes depend on *a)* how the policy in question affects the returns to factors; and *b)* farmers' ownership of those factors. In the case of developed OECD countries, a large share of the benefits of support that involves market interventions leaks to non-farming landowners and suppliers of purchased inputs (OECD, 2001). When there is a distribution of farm sizes, the benefits retained at the farm level also tend to be concentrated among larger (and wealthier) farmers.

There are several reasons why these results may not carry over seamlessly to poorer developing countries. Some specific features of developing country agriculture may result in policies being more effective than in a developed country context, others less.

One important feature of developing country agriculture is the joint role of the farm household as both a producer and consumer of food crops. This means that the effects of policies such as farm price support depend on what happens on both the supply side and the demand side. While higher prices stimulate production, they also raise the opportunity cost of consuming home-produced food. In many developing countries, an important share of farm households are net buyers of food, so raising farm prices could lower incomes for this group (unless they show a sufficient supply response to be transformed into net sellers).

A second factor is that many farm households confront high transaction costs when selling output or purchasing inputs. In the extreme, these transaction costs may be so high that the farmer remains cut off from the market altogether, producing only for home consumption (that is, subsistence). Under these circumstances a subsistence farm household may not benefit from higher farm prices, and could in fact lose via induced increases in land rental rates or in the prices paid for purchased inputs.

A third aspect is that rural households are heterogeneous in terms of their income sources, expenditure patterns and ownership of factors (particularly land), and will therefore be affected diversely by the direct and indirect impacts of policies. A comprehensive model of the agricultural sector in less-developed countries must consider the behaviour of structurally diverse agents, including commercial farms, semi-subsistence and subsistence farms, and landless rural households.

This chapter presents findings from a new model, the Development Evaluation Model (DEVPEM), which incorporates the above specificities. DEVPEM is a disaggregated model of the rural economy, which retains some key aspects of PEM, notably the imperfect transferability of land from one activity to another, which is central to the farmer's ability to respond to policy shocks. As with PEM, output and factor markets are linked, and the effects of policies on household incomes are determined by how those policies alter returns to factors that the household owns (land, labour and capital).

DEVPEM models are constructed for six countries, two in Africa (Ghana and Malawi); two in Asia (Bangladesh and Viet Nam); and two in Latin America (Guatemala and Nicaragua). The models are constructed using household level data from the FAO's Rural Income Generating Activities (RIGA) and market aggregates from the FAOSTAT database. The virtue of the RIGA database is that its variables have been standardised across countries, which makes it relatively straightforward to build and parameterise DEVPEM models, and helps ensure comparability across countries.

The six DEVPEM models are relatively stylised, and should not be considered as representing the full structural diversity of these countries or the precise way in which their rural economies function. Rather, the aim is a more modest one of shedding light on how basic structural differences among countries may affect agricultural policy outcomes. In particular, the model is used to investigate the impacts of market price support for food and cash crops, input subsidies and direct payments to households, as well as the impact of removing transaction costs which inhibit households' participation in markets. DEVPEM does not incorporate all market failures that may be important in low income countries (such as seasonal cash constraints and the absence of credit markets). It is a static model that is useful for examining the short- to medium-term impacts of agricultural policy interventions, but not the long-term effects, such as induced effects on investment decisions. These limitations constitute a potential avenue for further research and analysis.

3.2. The Development Policy Evaluation Model (DEVPEM)

The Development Policy Evaluation Model (DEVPEM) has been developed in order to analyse the welfare and distributional implications of alternative agricultural policies in developing countries. It is a rural economy model constructed by linking multiple farm household models in a general equilibrium framework, as in Taylor *et al.* (2005). A detailed motivation for the modelling approach is contained in Brooks, Dyer, and Taylor (2008), while a full exposition of the model with country-specific context is available in OECD working papers (OECD, 2011a; OECD, 2011b). This section outlines the main features of the model.

Farm households

The basic building block of DEVPEM is the agricultural household model, in which production, consumption and labour allocation decisions may be interdependent (Singh *et al.*, 1986). The household maximises its utility, which comes from consumption of home-produced goods, purchased goods, and leisure, subject to constraints on cash income, family time, endowments of fixed productive assets, and production technology. The household's consumption is constrained by its farm profits and income from marketed factors of production, such that the total value of goods consumed (from own production or purchased) is equal to the sum of all profits and the total market value of all endowments. If all prices are exogenous, the household can effectively make its production and consumption decisions sequentially, by first maximising its profit as a producer, given prices of outputs and inputs, and then using those profits to maximise utility, given prices of consumption goods. For each good, the difference between profit-maximising production and utility-maximising consumption is equal to net sales.

When prices are exogenous, the farm household's supply response is the same as that of a pure agricultural firm (as is implicitly the case in PEM). On the demand side, however, the household may decrease or increase its consumption depending on the relative strength of income and substitution effects. In economies in which farm households consume a large part of their output, market and welfare outcomes can differ widely from those predicted by an agricultural firm model. An increase in the price of an agricultural commodity may lead to an increase in production almost fully absorbed by a similar increase in consumption.

Transaction costs and missing markets

In developing countries, long distances to markets, a lack of infrastructure and imperfect information can result in high transaction costs, leading some households to remain self-sufficient in food production. In DEVPEM, transaction costs are captured via a price band, with households in remote areas selling on the market (or alternatively buying from the market) only if the difference between the market price and the self-sufficiency shadow price exceeds the associated transaction cost. When a household faces transaction costs for buying and selling a good, there will be an interval of prices (a price band) within which it uses its own subjective value of the good (*i.e.* the shadow price) to determine its optimal production and consumption levels. For prices within this band, separability between production and consumption decisions no longer holds.

DEVPEM explicitly models the effects of transaction costs and endogenous market participation. It is assumed that households located remotely from markets face proportional (or multiplicative) transaction costs when participating in markets. As buyers of consumption goods and inputs these households face an effective buying price that is higher than the market price, reducing their consumption possibilities. As producers, they face an effective selling price that is lower than the market price, creating a wedge between market and farm gate price.

Additional constraints, such as poorly functioning factor markets and seasonal liquidity constraints, may also impede the responses of households to market price changes. These constraints are not captured directly by DEVPEM. However, we assume fixed endowments, *i.e.* missing markets, for two factors of production: land and physical capital. These assumptions capture, indirectly, credit constraints that farmers may face, as well as the lack of well-functioning, formalised, land markets in the rural areas of many developing countries. When some factor markets are missing, the factor endowments of the household fundamentally determine its ability to produce and earn its living.

Imperfect land transferability

Many agricultural household models assume that land is a fixed input in each production activity. This assumption may be appropriate in the very short run, or when policies, customs or other considerations impede the smooth functioning of local land markets, as often is the case in developing countries. In the medium term, however, some land re-allocation of household land endowments across activities is likely to occur in response to various chocks. DEVPEM, like PEM, recognises that land may be transformable from one use to another, albeit imperfectly.

DEVPEM has three levels of land transformability, each with a different elasticity of transformation.[2] Figure 3.1 illustrates the principle. First, a distinction is made between land used for permanent cash crops and all other uses. Re-allocation of land at this level is assumed to be relatively difficult (indicated by the substitution parameter σ_1). At the second level of transformability, a distinction is made between pasture land and annual crops. Finally, at the third level, a distinction is made between food crops and annual cash crops, between which land is assumed to be relatively easier to re-allocate. Thus, σ_1 is smaller than σ_2, which in turn is smaller than σ_3.

Figure 3.1. **The three levels of land transformability in DEVPEM**

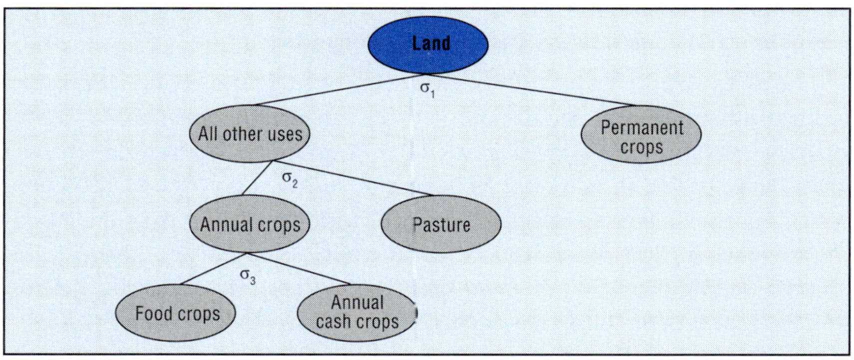

Note: The ease of land transformability from one use to another is expressed as the elasticity of transformation, σ. By assumption, $\sigma_1 < \sigma_2 < \sigma_3$, indicating that land re-allocation is easier the more similar any pair of agricultural activities are.

Household heterogeneity

Household heterogeneity is captured by partitioning all rural households into six separate categories, each with specific factor endowments, preferences, and production technologies. The purpose of this classification is to capture heterogeneity in the constraints that households face and their resulting responses to external shocks. The categories are rural non-farm households, large commercial farms, and small- and medium-sized farm households, with small and medium-sized farm households divided into "remote" and "non-remote" groups. "Remote" households confront prohibitive transaction costs, as defined above, which cut them off from food staple markets. Local markets often exist even in the most remote communities, but they operate in isolation from the rest of the world, and the prices on those markets reflect endogenous prices for the whole remote community. In that sense, the remote households in the model are not only representative of households in pure autarky, but also of households which buy and sell on local, isolated markets. They are pure subsistence households in the model baseline, but their market participation status may change.

Because the household distributions differ across countries in terms of land ownership and remoteness to markets, a farmer with, say, one hectare of cultivated land may be considered small in one country and fairly large in another. Thus, rather than defining category cut-off points as particular areas, we defined them as percentiles of the distributions, common for every country. This treatment of household categories has the advantage of providing a basis for cross-country comparison. Access to land was defined not in terms of land ownership but in terms of cultivated land. In some countries, access to land may be granted through traditional rights rather than formal land ownership, such that households cultivating land without a formal title should still be considered "owners" of that plot. Furthermore, access to land is likely to be relatively secure in the short to medium term, which is the relevant timeframe for DEVPEM simulations.

To define remoteness, we computed an index at the community level using distances to basic services or administrative centres (roads, buses, telephones, hospitals, schools, regional capitals, etc., variables depending on availability in each country survey). We considered as remote those households living in the 25% most remote communities according to this measure. Large farmers were not considered to be remote because we assume that their scale and the size of their assets would allow them to overcome transaction costs no matter where they might be located. It is well-established that land ownership is strongly correlated with market participation (Barrett and Dorosh, 1996), therefore it seems reasonable to assume that all large farmers are connected to markets. In DEVPEM, small- and medium-sized farms in remote areas are considered to be subsistence producers with respect to food crops in all six countries.

Rural economy linkages

The six representative household groups in DEVPEM constitute the rural economy. The rural economy is, in turn, linked to the urban economy, or "the rest of the world". Opening up for general-equilibrium effects implies taking into account the markets for inputs and factors of production, as well as the effects on other goods markets. For example, price support for maize may increase labour demand, inducing a higher wage rate in the rural economy. This might benefit landless households (whose only income source is wage labour) who – if maize is only a small share of their household budget – might ultimately gain more as wage labourers than they lose as maize consumers. While general-equilibrium linkages in DEVPEM are more involved than this, the above example illustrates the potential strength of general equilibrium analysis. Establishing the direction and quantitative impacts of policy shocks cannot be done analytically but requires a programming approach.

The magnitude of these linkage effects is limited, however, by the fact that most prices in DEVPEM are assumed to be exogenous to the rural economy. There are seven commodities defined in the model: six agricultural commodities, all produced in the rural economy, and one composite non-agricultural good, representing all consumption goods produced outside the rural economy. There are five factors of production in the model: family labour, hired labour, land, physical capital, and intermediate inputs (such as seeds and fertiliser). All commodity prices are assumed to be exogenous, that is, determined in the urban economy or on the world market. In terms of production factors, the wage rate and the price of intermediate inputs are determined endogenously in the rural economy, although we consider an alternative scenario in which intermediate inputs, such as fertiliser, are supplied exogenously at the world price. Since markets for land and physical capital are assumed to be missing, there are no market prices for those factors. The elements and defining characteristics of the DEVPEM model are summarised in Table 3.1. For a detailed explanation, the reader is referred to the technical documentation of the model in OECD (2011a and b).

Data sources and model calibration

The basic data platform for DEVPEM consists of disaggregated social accounting matrices, constructed with household survey data compiled by the United Nations Food and Agricultural Organisation in its Rural Income Generation Activities (RIGA) database and with data from the FAOSTAT database.[3]

The model consists of a set of variables (for which we have observations) and a set of relationships among variables, defined by equations with parameters (for most of which we do not have observations). In order to make

Table 3.1. **DEVPEM overview: household groups, production factors, and commodities**

Households groups	Defining characteristic or assumption
Non-farm households	Landless rural households, no farm income.
Small remote farm households	Face transaction costs for buying and selling farm goods.
Small non-remote farm households	Small land holdings, no transaction costs.
Medium-sized remote farm households	Face transaction costs for buying and selling farm goods.
Medium-sized non-remote farm households	Medium land holdings, no transaction costs.
Large farm households	Relatively large land holdings, no transaction costs.
Factors of production	**Defining characteristic or assumption**
Family labour	Used on the own farm or supplied on the rural labour market; wage rate determined endogenously.
Hired labour	Wage rate determined endogenously; not all farms hire labour.
Land	Missing market; imperfect transferability from one use to another.
Physical capital	Missing market.
Intermediate inputs (seeds, fertiliser, etc.)	Supplied by the urban economy at an increasing marginal cost.
Commodities[1]	**Defining characteristic or assumption**
Main cash crop	Produced but not consumed by rural households; price determined exogenously.
Main food staple Other food staples Other annual crops Other cash crops Livestock products	Produced and consumed (partly) by rural households; price determined exogenously.
Non-agricultural ("market") goods	Produced by the urban or foreign economy and consumed (partly) by rural households. Price determined exogenously.

1. Classification varies somewhat by country model.

the model operational and tractable, we must calibrate it, that is, find the missing parameter values, using actual production and consumption data for each country for which the model is applied. The aim of calibration procedures is to find parameter values such that the observed data represent a solution to the model. Our calibration procedure is based on a social accounting matrix (SAM) for each of the six countries. A SAM provides a picture of all flows of money and goods in an economy in matrix form, where rows represent the incomes of economic actors and columns represent expenditures. An advantage of using a SAM is that, by construction, all cash constraints and market clearing conditions are satisfied for all accounts in the matrix, which is consistent with general equilibrium theory. This is why computable general equilibrium (CGE) practitioners often parameterise models using SAMs. Each country SAM has the same structure, even if specific activities and goods differ between countries. Keeping the structure the same for each country facilitates comparability across countries and simplifies extensions of the model, with applications on additional countries.

3.3. Agricultural policy simulations with DEVPEM

In this section we report the results of policy simulations using DEVPEM. We analyse the effects of five different policies in each of the six countries included in the study. Three of the policies are *market interventions*, in the form of market price support, a production subsidy, and an input subsidy; one of them is a *social transfer*, in the form of an unconditional cash transfer; and one is a *public-good investment* that lowers transaction costs for remote households and facilitates access to markets. We are interested primarily in the ability of each policy to increase the welfare of rural households, how costs and benefits are distributed across household groups in each country, and how cost efficient each policy is in terms of raising the welfare of the targeted population for every dollar spent on the policy. It is important to bear in mind that we are concerned with a particular set of costs and benefits that DEVPEM is designed to capture. There may be other costs, such as administrative costs and losses due to corruption. Equally, there could be dynamic effects that are beyond the scope of the current model.

To the extent that agricultural policies are aimed primarily at increasing the welfare of rural households, market-intervening policies are second-best, in the sense that they involve deadweight losses and leakages to unintended beneficiaries. Public-goods investments do not suffer from deadweight losses, in the theoretical sense, but from the standpoint of raising income, they may be difficult to target and generate long-term benefits that are outside the scope of the DEVPEM model. From a targeting and efficiency point of view, direct social cash transfers to specific groups are therefore better than market intervention and public-goods investments. Before analysing the results of the policy simulations, we discuss the ways in which the policy experiments are implemented in the model.

Policy implementation

The market price support (MPS) and production subsidy (PS) experiments are both targeted at agricultural commodity markets, the main difference between the two policies being that the former affects consumer prices while the latter does not. Production subsidies, which are formally equivalent to a deficiency payment equal to the difference between a target support price and the market price, are rarely implemented in developing countries, as they necessitate the use of scarce budgetary resources. However, they provide an instructive comparison with MPS policies because of this basic difference in effect. Input subsidy (IS) policies consist of interventions in markets where farmers are buyers and consumers do not participate, such as the markets for seeds and fertiliser. Common to all policy experiments is the assumption that the urban economy (urban consumers and taxpayers) bears all the explicit costs of the policies in terms of taxes. Some of the policies also imply implicit costs to the urban economy in terms of consumer surplus losses due to higher commodity prices.

Market price support

The MPS policy is implemented as a price floor, or a regulated minimum price, for the targeted commodity. It raises the price above the world market price for farmers and rural consumers, as well as for urban consumers. In the rural economy, farm households gain as producers and lose as consumers, their net gain depending on their production surplus. As long as they produce more than they consume they are likely to gain from the policy.

Implementing MPS will cause the quantity demanded of the targeted commodity to fall and the quantity supplied to increase. If the country is initially a net exporter of the commodity, MPS will increase the domestic excess supply and the government will have to use taxes to cover the difference between the targeted price and the world market price, taking the surplus into storage or selling with the use of implicit export subsidies on the world market. If the country is a net importer of the targeted commodity, the government can impose an import tariff equal to the difference between the target price and the world market price in order to keep the domestic price at the target level. In this case, households in the urban economy lose in terms of higher prices of the commodity (a consumer surplus loss).

We analyse the effects of market price support for three agricultural commodities: the main food crop, the main cash crop, and livestock products. As indicated in Table 3.1, farm households consume some of the food crops and some of the livestock products they produce. The main cash crop, however, is produced for "export" only, either for the world market of for the urban market. The experiment consists of raising the domestic price 10% above the world market price of one commodity at the time. This price change is small enough to assume that the model parameters for consumer preferences and production technology remain valid, yet large enough to cause noticeable behavioural adjustments among households.

Production subsidy

In contrast with MPS, a production subsidy (PS) does not affect consumers in terms of higher prices. This follows from the assumption of exogenous output prices. The subsidy, as implemented here, gives the farm household a mark-up on the world market price for each unit of the commodity it produces and sells. To the extent that the farm household consumes the targeted commodity, it is able to buy it in the market at the world market price. This means that, in the rural economy, the quantity supplied of the good increases due to the higher seller price, but the quantity demanded remains unchanged. Consequently, the rural excess supply of the commodity increases, although by less than under MPS. Urban consumers face a tax cost under this policy but are unaffected as buyers of the good. The tax needed to finance the policy amounts to the domestic quantity produced times the subsidy amount per unit of output. As with MPS, the domestic surplus of the commodity will be exported at the world market price. Designed in this way, a production subsidy has an advantage over market price support in that it leaves rural households unaffected on the consumer side. It may, however, be difficult to administer, owing to the need for farmers to provide sales receipts in order to receive payments.

As with the MPS experiments, we analyse the effects of a production subsidy for the main food crop, main cash crop, and for livestock products. Since rural households are assumed not to consume any of the cash crops, the effects on the rural economy of an MPS and PS are identical for cash crops. As in the MPS experiments, we assume that the subsidy consists of a 10% mark-up on the world market price.

Input subsidies

An input subsidy enables farm households to buy intermediate inputs at a lower price than the market price. While there are various ways of implementing such a policy, we assume here that the policy is implemented as vouchers given to farmers, effectively giving them a 10% discount on the targeted input. The effect of subsidising an agricultural input depends crucially on the market structure for that input. The extent to which the price paid by farmers will actually fall depends on how sensitive the supply of the input is to price changes. The more price-elastic the supply, the more effective the subsidy will be in lowering the price paid by farmers. In the case of an exogenous price (determined by the world market), the price paid by the farmer will fall by the entire amount of the voucher and there will be no leakage to the suppliers of the input. In the other extreme – the case of fixed supply – the price paid by farmers for the input will not fall at all; the entire value of the voucher will be passed on to the suppliers of the input.

If the market for intermediate inputs is well connected with the world market, the price is essentially exogenous and an input subsidy will induce little or no leakages to suppliers. If, however, these inputs are supplied at an increasing marginal cost, for example due to limited production capacity, then some degree of leakage is inevitable. Given that it is difficult to be sure of the efficiency of domestic distribution channels, we analyse the effects of an input subsidy under two scenarios. The benchmark assumption is that the supply of inputs is relatively elastic.[4] The alternative scenario is that the input market is perfectly connected with the rest of the world so that the price is fully exogenous at the going world market price. Since consumers do not to participate in the markets for intermediate inputs used in agricultural production, they suffer no losses in terms of consumer surplus. The cost of the policy is borne solely by urban taxpayers and amounts to the total quantity sold of the targeted input times the voucher value.

Cash transfers

The cash transfer that we consider in this chapter is an agricultural policy only in the sense that it is targeted to farm households. In contrast with policies that involve market interventions, it is not tied to crop production or to a specific use of inputs. Theoretically, a cash transfer is the best way of raising the welfare of specific household groups since it is free of leakages and does not distort markets incentives. Moreover, it can in principle be targeted to poorer households. The use of any other instrument is therefore likely to reflect either administrative challenges or the simultaneous pursuit of other policy objectives, as discussed in the introduction. The way we simulate the cash transfer in this chapter is essentially trivial and used only as a benchmark policy experiment. The policy is untargeted and each farm household group receives a cash transfer equal to one per cent of household income.

A public-goods investment

The notion of public goods is used here in a broad sense to distinguish goods that benefit a large group of people (such as roads, irrigation systems, public health centres, and schools) from "private" goods (such as consumer goods and agricultural production equipment) that only benefit their sole owner. Like unconditional cash transfers, rural public-goods investments are an agricultural policy only in the loose sense that they benefit people of whom a majority are farmers.

Investments in public goods can take many forms. The type that we consider here is an infrastructure-type of investment that reduces transaction costs for remote households. This does not have to be related to transportation. It could, for example, be an investment in a mobile telecommunication network, connecting people living in remote areas to people in towns and cities, allowing them to receive information about current prices of crop or livestock products or current labour demands in different regions. The policy in our experiment is assumed to be designed in a way that, after the investment, remote households face the same set of effective prices as other (non-remote) households. This policy only affects two of the household groups and we cannot say anything specific about the cost of the policy or its relative efficiency.

Main findings from agricultural policy simulations

The welfare effects of the above experiments are shown in Table 3.2. We briefly summarise the main findings here, referring the reader to OECD (2011a) for a more comprehensive discussion of the results.

Market price support

MPS for food crops harms rural non-farm households in five of the six countries, with consumption side losses exceeding any gains from higher wage incomes. The benefits to farm households are typically small, and are concentrated among medium-sized and large farmers. Only in Ghana are there significant benefits across all five farm household groups. MPS for livestock products similarly harms rural non-farm households in all six countries. Small farm households are either unaffected or suffer a small welfare loss in all countries but Nicaragua, where they gain significantly. In the two African countries and in Viet Nam effectively no one benefits from the policy. Gains are small in Bangladesh and Guatemala, the only case in which MPS for livestock significantly increases welfare being for medium-sized and large farmers in Nicaragua, a consequence of the large share of livestock in the product mix of Nicaraguan farmers. By contrast, there are no significant welfare losses in the rural economy from MPS for cash crops, because there is no domestic consumption. Except in Ghana, where small farmers grow cocoa beans, the welfare effects are negligible for small farmers and non-farm rural households. The gains are significant for medium and large-scale farmers in three of the six countries: Guatemala, Nicaragua and Viet Nam.

Production subsidies

A production subsidy for the main food crop, which has the same economic effect as a deficiency payment, affects farmers on the producer side in the same way as market price support, but leaves the consumer side unaffected. As a result, no household group loses significantly, although the gains tend to be concentrated among medium and large-scale farmers more than small farmers.

Input subsidies

In contrast with MPS, an input subsidy (IS), modelled as a discount on the price farmers pay for intermediate inputs, benefits "remote" households who are cut off from output markets. Welfare increases are higher for medium and large scale farmers, but significant for small farmers in all countries except Bangladesh. The gains are largest in Viet Nam, where existing use of fertiliser is highest. The input subsidy experiment is run under the assumption that the price of inputs is endogenously determined. If input prices are exogenous, *i.e.* set at the world market level plus a fixed per unit distribution cost, then the gains approximately double for each household group.[5]

Table 3.2. **Simulation results of rural household welfare effects of various agricultural policies (% change)**

	Ghana	Malawi	Guatemala	Nicaragua	Bangladesh	Viet Nam
Market price support, food staple						
Non-farm	–0.11	–1.71	–0.25	0.31	–0.81	–0.89
Small remote	2.60	–0.08	–0.03	0.29	–0.04	0.30
Small non-remote	1.29	–0.93	0.03	0.49	–0.06	0.30
Medium remote	0.68	–0.13	–0.04	0.16	–0.44	–0.51
Medium non-remote	2.32	0.25	0.26	0.63	0.91	1.46
Large farm	2.29	1.22	0.40	0.28	1.02	1.05
Market price support, cash crop						
Non-farm	0.34	0.00	0.23	0.44	0.01	0.15
Small remote	1.26	–0.08	0.62	0.68	0.06	0.13
Small non-remote	0.11	–0.04	0.25	0.34	0.07	0.20
Medium remote	1.15	0.59	2.70	2.04	0.16	1.04
Medium non-remote	0.47	0.44	1.78	2.20	0.26	1.14
Large farm	0.78	0.12	1.77	2.18	0.27	5.09
Market price support, livestock						
Non-farm	–0.34	–1.37	–1.13	–0.21	–0.56	–1.36
Small remote	–0.03	–0.01	–0.02	2.31	–0.15	–0.67
Small non-remote	0.04	–0.38	–0.32	0.92	0.07	–0.38
Medium remote	0.04	–0.01	–0.01	3.93	–0.22	–0.53
Medium non-remote	0.15	0.05	0.76	1.89	0.53	0.13
Large farm	0.02	0.23	1.49	3.67	0.61	0.16
Production subsidy, food staple						
Non-farm	1.68	0.01	0.04	0.42	0.12	0.32
Small remote	2.60	–0.08	–0.03	0.29	–0.04	0.30
Small non-remote	3.04	0.67	0.38	0.59	0.60	1.21
Medium remote	0.68	–0.13	–0.04	0.16	–0.44	–0.51
Medium non-remote	4.08	1.13	0.56	0.71	1.66	2.32
Large farm	3.68	1.74	0.57	0.35	1.57	1.82
Input subsidy						
Non-farm	0.28	0.14	0.03	0.13	0.03	0.41
Small remote	0.77	0.56	0.56	0.26	0.16	1.61
Small non-remote	0.46	0.53	0.10	0.22	0.19	0.85
Medium remote	1.25	0.80	0.72	0.59	0.62	2.54
Medium non-remote	0.88	0.88	0.88	0.67	0.56	1.77
Large farm	1.14	1.03	1.20	0.68	0.37	2.80

Table 3.2. **Simulation results of rural household welfare effects of various agricultural policies (% change)** *(cont.)*

	Ghana	Malawi	Guatemala	Nicaragua	Bangladesh	Viet Nam
Public-good investment						
Non-farm	0.2	0.0	0.1	0.4	0.0	0.1
Small remote	1.2	0.1	1.3	3.5	0.1	1.2
Small non-remote	0.1	0.0	0.0	0.1	0.0	0.1
Medium remote	3.5	0.7	4.4	7.2	0.2	5.3
Medium non-remote	0.0	0.0	–0.1	–0.1	0.0	–0.2
Large farm	–0.1	–0.1	–0.2	–1.1	0.0	–0.5

Note: The market price support simulations assume a 10% exogenous price increase in the targeted commodity. The production subsidy simulations assume that farmers are given a 10% subsidy for all output produced of the targeted commodity. The input subsidy simulation assumes that farmers are given vouchers, which gives them 10% discount on the targeted agricultural input. The public-good investment simulation assumes that transaction costs are eliminated for households located remotely from markets.

Source: DEVPEM simulation results, Annex 1, "Modelling the Distributional Implications of Agricultural Policies in Developing Countries: The Development Policy Evaluation Model (DEVPEM)".

Cash transfers

When households receive an unconditional transfer equal to 1% of their original income, they also observe a 1% increase in their final welfare, with general equilibrium effects (*e.g.* as a result of households switching between activities) not significant.

We measure the cost efficiency of alternative instruments as the overall change in rural welfare divided by the cost to urban taxpayers and (in the case of food price support) consumers. The average efficiency across all six countries is shown in Figure 3.2. The benchmark experiment of an unconditional cash

Figure 3.2. **Cost efficiency of various agricultural policies (averaged across the six countries)**

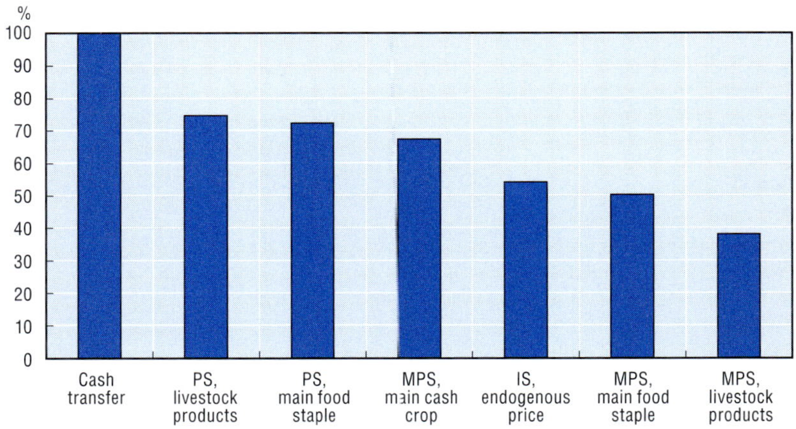

Source: OECD (2011b).

StatLink http://dx.doi.org/10.1787/888932552665

transfer, as designed, has a cost efficiency of 100%. Among policies that involve interventions in commodity markets, those without negative consumer side-effects have higher levels of efficiency. The efficiency of the production subsidy exceeds 70%, while the efficiency of MPS to cash crops is on average 67%. The policies with the lowest levels of efficiency are those that harm rural households on the consumption side, namely price supports for food crops and livestock products, with efficiency levels below 50% in many cases. The efficiency of input subsidies is low (54%) on the assumption that prices are determined endogenously, but much higher if prices are taken to be exogenous (over 90%).

3.4. Conclusions

The results of the DEVPEM simulations show that, for alternative policy instruments, there are significant differences in household-level impacts across countries. The observed effects also differ in some systemic ways with those found in OECD countries. Some general findings stand out.

The first conclusion from these policy experiments is that no untargeted agricultural policy intervention is pro-poor within the rural economy. Market price support for food crops harms net buyers of food, often the poorest farm and non-farm (landless) households, although the proportion of net buyers varies significantly across countries.[6] Support for cash crops does not have this drawback; however, cash crops are typically grown by farmers with relatively high incomes, so support seldom reaches the incomes of the poorest.

While all agricultural policy instruments are less efficient than a simple cash payment (in terms of net rural benefits relative to urban cost), the degree of inefficiency of some market interventions, notably input subsidies, is not inevitably as high as observed in developed OECD countries. The reason is that farmers are less likely to rent land than in OECD countries and they tend to purchase fewer inputs, which in turn imply fewer leakages away from the farm level. This result may have implications if agricultural policies can be targeted and if other factors germane to policy design and implementation are sufficiently important.

Given that a cash transfer is the most efficient instrument, and can in principle be targeted to low income households, the justification for using an alternative instrument must come from some effect which is not captured by the DEVPEM model. One is the possibility of administrative difficulties with cash-based programmes, for example due to the absence of a population registry, or because of concerns about corruption (although these concerns also apply to other instruments). Another is the possibility of dynamic gains from market interventions, in terms of enabling farmers to overcome market failures and break out of "poverty traps". The latter has been invoked as a justification for input subsidies (Dorward, 2009).

Input subsidies tend to benefit those farmers who are using inputs already – often larger farmers. The extent to which the benefits of support are retained by farmers depends partly on the degree to which the price of inputs rises in response to the increase in input demand. Parallel investments that increase distribution capacity and help keep marginal distribution costs constant would reduce the leakage of benefits away from the farmer. Input subsidies can also in principle be targeted to improve their distributional outcomes. A wide range of other pros and cons are relevant when considering whether input subsidies are a relevant policy option – these are taken up in Chapter 5.

A significant finding is that a public investment which reduces transaction costs is the only pro-poor instrument, since it helps remote farm households, who are typically poorer than those engaged with markets. However, we do not know the cost of generating these improvements. If those investments bring down marginal input distribution costs, they could also improve the effectiveness of input subsidies. The implication here is that, if input subsidies are to be used, they should be accompanied by investments in infrastructure. Moreover, weak infrastructure, which impedes market profitability, is often a reason for farmers failing to use inputs in the first place – so such investment could replace as well as complement an input subsidy.

In general, the results show that direct payments are the most efficient way of boosting incomes in the short term, while public investments, which should also have broader long-term pay-offs, have short- to medium-term impacts that are pro-poor. These results are therefore fully consistent with the best practice advice established for OECD countries of using social policies to safeguard incomes in the short term, and non-distorting public investments to boost competitiveness in the long term.

Whether the advice provided to OECD countries needs to be qualified depends partly on whether considerations outside the model are sufficiently important to offset the shortcomings of market interventions on both efficiency and equity criteria. Some of these caveats are taken up in Chapter 4, which considers whether there is a role for government in stabilising markets, and Chapter 5, which considers the arguments for and against using input subsidies. Other features of developing country agriculture that DEVPEM is currently unable to capture, such as liquidity or risk constraints, could be addressed by further development of the model. Such a development could provide insights into a wider range of policy impacts.

Notes

1. The authors would like to thank Alberto Zezza and Katia Covarrubias for their help in providing the RIGA data upon which this analysis is based. They are also grateful for comments received at the PEM Experts' Group meeting on 24-25 September 2009, when the DEVPEM model was in the process of construction.
2. The elasticity of transformation is defined as the percentage increase in land use in one activity, given a 1% decrease in land use in the other activity. Land transformability across activities is modelled in DEVPEM with a constant elasticity of transformation (CET) function.
3. Davis *et al.* (2010) and Winters *et al.* (2009) describe the RIGA database and present detailed insights on asset holdings and income sources among rural households, based on these data. Further information is available at *www.fao.org/economic/riga*.
4. We assume rising internal distribution costs, reflecting infrastructure bottlenecks, which are reflected in a supply elasticity of 2.0. Quizón and Binswanger (1986) in an application to India assume a higher elasticity of 4.0, based on openness to trade. Ryan and Perrin (1974) in an examination of the market for potatoes in Peru argue that it is plausible to assume perfectly elastic supplies, since fertiliser is imported. We adopt the latter as an alternative assumption.
5. The nature of the domestic market for fertiliser will be context specific. We assume rising internal distribution costs, reflecting infrastructure bottlenecks, which are reflected in a supply elasticity of 2.0. Quizón and Binswanger (1986) in an application to India assume a higher elasticity of 4.0, based on openness to trade. Ryan and Perrin (1974) in an examination of the market for potatoes in Peru argue that it is plausible to assume perfectly elastic supplies, since fertiliser is imported. We adopt the latter as an alternative assumption.
6. In an analysis of eleven countries for which RIGA data are available, Zezza *et al.* (2008) find that rural net buyers of food outnumber rural net sellers in all but one country (Viet Nam).

References

Barrett, C., and P. Dorosh (1996), "Farmers' welfare and changing food prices: Nonparametric evidence from rice in Madagascar", *American Journal of Agricultural Economics*, Vol. 78, pp. 656-669.

Brooks, J., G. Dyer and J.E. Taylor (2008), "Modelling agricultural policy impacts in less developed countries", *OECD Food, Agriculture and Fisheries Working Papers*, No. 11, OECD, Paris.

Davis, B., P. Winters, G. Carletto, K. Covarrubias, E. Quinones, A. Zezza *et al.* (2010), "A cross country comparison of rural income generating activities", *World Development*, Vol. 38, pp. 48-63.

Demeke, M., G. Pangrazio and M. Maetz (2008), *Country responses to the food security crisis: nature and preliminary implications of the policies pursued*, Food and Agriculture Organisation of the United Nations, Rome.

Dorward, A. (2009), "Rethinking agricultural input subsidies in a changing world", paper prepared for the Food and Agriculture Organisation of the United Nations, Rome.

Jones, D. and A. Kwiecinski (2010), "Policy responses in emerging economies to international agricultural commodity price surges", *OECD Food, Agriculture and Fisheries Working Papers*, No. 34, OECD, Paris, doi: *10.1787/5km6c61fv40w-en*.

OECD (2001), *Market Effects of Crop Support Measures*, OECD, Paris.

OECD (2002), *Agricultural Policies in OECD Countries: A Positive Reform Agenda*, OECD, Paris.

OECD (2005), *The Six-Commodity PEM Model: Preliminary Results*, OECD, Paris.

OECD (2011a), "Modelling the distributional implications of agricultural policies in developing countries – the Development Policy Evaluation Model (DEVPEM)", *OECD Food, Agriculture and Fisheries Working Papers*, No. 50, OECD, Paris.

OECD (2011b), "The Development Policy Evaluation Model (DEVPEM): technical documentation", *OECD Food, Agriculture and Fisheries Working Papers*, No. 51, OECD, Paris.

Ryan, J.G. and R.K. Perrin (1974), "Fertilizer response information and income gains: The case of potatoes in Peru", *American Journal of Agricultural Economics*, Vol. 56, pp. 337-343.

Singh, I., L. Squire, and J. Strauss (eds.) (1986), *Agricultural household models*, The Johns Hopkins University Press for the World Bank, Baltimore, MD.

Taylor, J.E., G. Dyer and A. Yúnez-Naude (2005), "Disaggregated rural economy wide models for policy analysis", *World Development*, Vol. 33, pp. 1671-1688.

Quizón, J. and H. Binswanger (1986), "Modelling the impact of agricultural growth and government policy on income distribution in India", *The World Bank Economic Review*, Vol. 1, pp. 103-148.

Winters, P., B. Davis, G. Carletto, K. Covarrubias, E. Quinones, A. Zezza *et al.* (2009), Assets, activities and rural income generation: evidence from a multicountry analysis, *World Development*, Vol. 37(9), pp. 1435-1452.

Chapter 4

Stabilisation Policies in Developing Countries after the 2007-08 Food Crisis

by
Phil Abbott[1]

Conventional best practice advice for risk management strategies tends to focus on long-run agricultural development, trade liberalisation, the provision of safety nets and private market solutions to risk. However, if world price spikes like those observed in 2008 are an infrequent but real event, policy recommendations need to take into account the greater prevalence of market failures in many developing countries and associated underdevelopment of marketing institutions. While policy should rely on liberal trade in most years, a short-run stocks policy may be a viable option, due to delays in import arrival, imperfect information on the harvest, and inter-seasonal price dynamics. Moreover, trade policy adjustments are likely to be perceived as necessary when infrequent world price spikes reoccur. The challenge to implementing such policies lies in ensuring consistent, predictable and transparent governance so that interventions make outcomes better, not worse.

4.1. Introduction

After a prolonged period of stability starting about 1998, international grain prices began to rise in 2007 and spiked in 2008 (von Braun, 2008). Between October 2006 and January 2008, world wheat prices rose 74%, while world rice and maize prices increased 27% and 45%, respectively. Wheat prices rose another 19% to their peak in March 2008 and maize prices rose another 39% from January 2008 levels, peaking in June. The most spectacular rise was for rice, as world rice prices increased another 157% by April 2008. Figure 4.1 shows the relative stability of these prices between 1998 and 2006 and then the dramatic price increases in 2007 and especially 2008 (IMF, 2010).

The FAO (2008b) and USDA (Rosen *et al.*, 2008) have estimated that between 75 and 133 million additional poor people suffered from hunger and malnutrition as a result of these world price increases. The World Bank (2008a) estimated that an additional 105 million people suffered extreme poverty. Initial expectations were that high prices would persist (OECD and FAO, 2008), but global recession reduced the spikes (Abbott, Tyner and Hurt, 2008 and 2009). Price levels remain elevated relative to earlier this decade, global recession means poverty and hunger problems persist, and many believe we are now in an era of more volatile food prices (von Braun, 2009; Delgado and Townsend, 2009). Food policy changed in many countries to counteract these outcomes. While the food crisis has abated, food security issues persist and food policies need to be re-examined.

Policy responses to world food price spikes

This food crisis of 2007-08 brought substantial responses by national governments and by the international donor community to address poverty and hunger, to renew efforts aimed at increasing agricultural productivity, and to protect consumers broadly. While the international community has focused on safety nets in the short run and fostering growth in agricultural production in the medium to long run (UNHLTF, 2008; Viatte *et al.*, 2009), national governments of developing countries pursued a number of policies to stabilise domestic markets and to isolate their consumers from events in world grain markets (Abbott, 2009). An FAO study (Demeke, Pangrazio and Maetz, 2009), which examined policy responses in 81 developing countries, found that 43 countries reduced tariffs and 25 countries imposed export taxes or restrictions to mitigate the effects of higher international prices. Table 4.1

Figure 4.1. **International grain price indices**

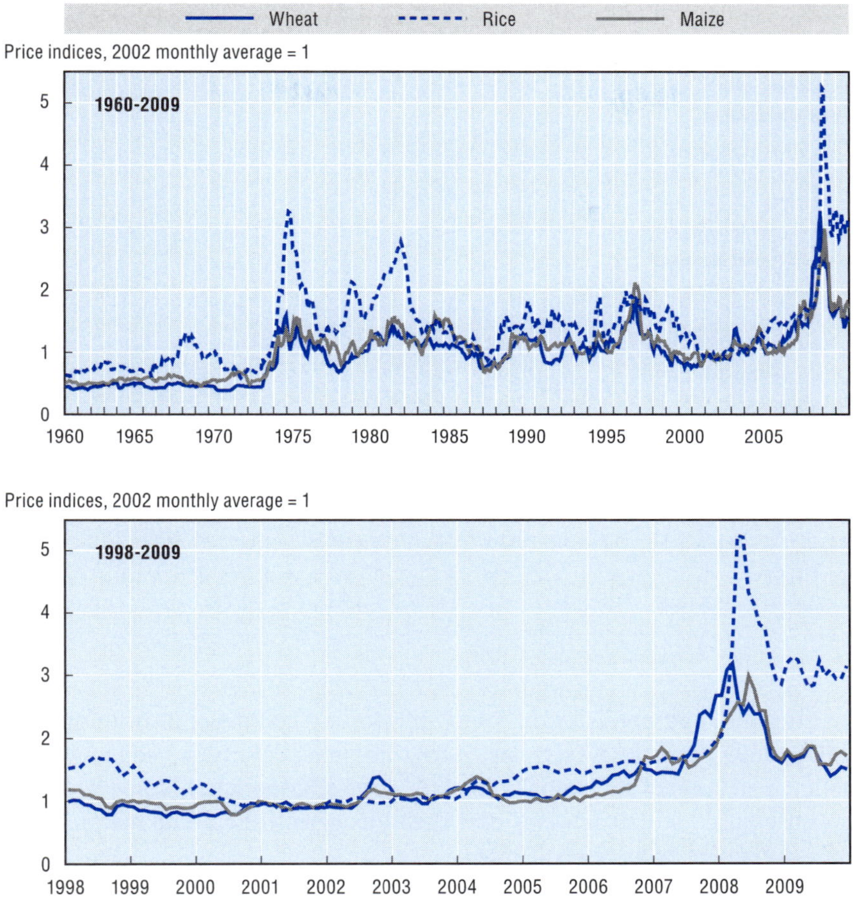

Source: IMF (2010), *International Financial Statistics.*

StatLink http://dx.doi.org/10.1787/888932552684

shows that the FAO also found domestic measures (*e.g.* cutting taxes on food, subsidies, and stocks releases) used to complement these trade measures aimed at preventing transmission of world price variability to domestic markets. Domestic political objectives, including concern for urban consumers and quelling food riots, took precedence over exacerbating instability in international markets (Wodon and Zaman, 2008). In the case of rice, several authors (*e.g.* Timmer, 2008a and b; Dawe, 2008; Diouf, 2008) highlighted the importance of export bans by key traders leading to the especially strong spike in world rice prices. Thus, the food crisis set a new stage for stabilisation policy in developing countries, in which trade liberalisation looked to be an ineffective option and so was rejected by many governments. Government stabilisation goals, and even public marketing

Table 4.1. **Trade based policy measures commonly adopted (as of 1 December 2008)**

	Africa	Asia	Latin America	Overall
Countries surveyed	33	26	22	81
Market interventions				
Trade policy				
Reduction of tariffs and customs fees on imports	18	13	12	43
Restricted or banned export	8	13	4	25
Domestic market measures				
Suspension/reduction of VAT or other taxes	14	5	4	23
Released stocks at subsidised prices	13	15	7	35
Administered prices	10	6	5	21
Production support				
Production support	12	11	12	35
Production safety nets	6	4	5	15
Fertiliser and seed programmes	4	2	3	9
Market interventions	4	9	2	15
Consumer safety nets				
Cash transfers	6	8	9	23
Increase disposable income	4	8	4	16

Source: Demeke, Pangrazio and Maetz, 2008.

institutions, had persisted in many developing countries (Cummings and Gulati, 2009; Jayne and Tschirley, 2009), but some countries used new trade policy instruments to (partially) isolate domestic markets. Moreover, many countries used stocks releases that theory suggests should be ineffective at changing prices in an open economy.

Policy responses by developing country governments often constituted a reversion to past policy regimes, or at least past policy objectives. Structural adjustment programs of the IMF and World Bank had reduced the role of the state in grain markets in many instances, and had encouraged a greater role for the private sector (Abbott, Andersen and Tarp, 2010). Continuing desire for stable domestic markets led the World Bank to identify best practices for risk management in agriculture in a privatised market setting (Byerlee, Jayne and Myers, 2005) and to work with developing countries to adopt new market based institutions (CRMG, 2008). They encouraged establishing and using private risk management institutions including futures markets, crop insurance and forward pricing, and argued that trade liberalisation was an important component of that strategy. They downplayed the role of public stockholding, and of policies pursued by governments to stabilise domestic prices. Galtier (2009a, b) argued that this approach failed in the face of the 2007-08 food crisis, and that a new strategy with a greater role for government is called for. Countries had not implemented to any significant extent the private sector risk management strategies advocated by the World Bank (CRMG, 2008; Galtier, 2009). The trade and domestic policies adopted in the face of the crisis (Demeke, Pangrazio and Maetz, 2009) served to isolate markets, and were often the kinds of policies used in the pre-structural adjustment era. They reflected a set of national objectives that favoured domestic price stabilisation and were less concerned with implications for greater instability in international markets that such policy responses would bring (Abbott, 2009).

Stabilisation policy debates

Subsequent debate has looked at both appropriate trade and domestic policy for developing countries, and at initiatives to stabilise international markets. Sarris (2009) has proposed both greater use of futures markets by developing countries, and establishing an international clearing house along the lines used by commodity exchanges to insure contracts are honoured in world markets and world market supplies are reliable. Von Braun and Torero (2009) have proposed an international virtual reserves scheme to combat world price spikes. Galtier (2009a, b) argued for a complementary combination of domestic trade and stocks policies to preserve stability in domestic markets. Timmer (2008a, b) has argued that it is unrealistic to expect Asian rice traders to back away from the public stabilisation schemes that kept world price spikes out of their domestic markets.

Substantial bodies of literature have examined domestic stabilisation policy in developing countries, international alternatives to those regimes, and private risk management alternatives to public stabilisation initiatives. Anton (2009) has argued that a holistic approach to risk management is needed, where stabilisation policy takes into account private market strategies, rather than pursuing piecemeal price or income interventions. Numerous studies have examined what happened in 2007-08 in specific developing countries and how policies and markets coped with the world price increases (see Abbott, 2009 for a review), emphasising the extent to which price stabilisation objectives were pursued. Weak institutional development and preference for price stability have prevented broad-based, holistic approaches to risk being pursued extensively in most developing countries.

Past literature has been critical of public stockholding and stabilisation policy managed via stocks or trade policy. Seminal work by Newberry and Stiglitz (1981) argued that international stockpiling and price stabilisation would be ineffective and costly in stabilising producer welfare. The various international commodity agreements were judged to be failures on both economic and political grounds (Sarris, 1998; Gilbert, 1996). Stockpiles tended to be held for long periods, due to the asymmetric nature of world price and domestic production distributions, leading to high stabilisation costs. Moreover, especially for international schemes, parties to commodity agreements did not agree on the basic objectives of those institutions. As early as 1984 financial alternatives were being explored due to the perceived high transaction costs of stockpiling alternatives (Huddleston *et al.*, 1984). Work persisted in looking at international stabilisation schemes until the early 1990s, due to the persistent interest in stabilisation in developing countries (*e.g.* Braverman *et al.*, 1992; Abbott *et al.*, 1993). After that time, two strands of literature have dominated. One has focused on risk management strategies in a private market setting (see Byerlee, Jayne and Myers, 2005 and Anton, 2009 for reviews). Another is the considerable work examining agricultural policies in developing countries, where issues relevant to stabilisation policy and the role of the state in agricultural markets have remained relevant (*e.g.* Jayne and Tschirley, 2009).

Roadmap

This chapter will examine stabilisation policy from the perspective of a developing country government, focusing on domestic policy alternatives and implications for trade policy. The key question is whether developing country governments should intervene to stabilise their domestic markets, and if so, what policy instruments and institutions should they employ? In particular what (if any) role should public stockpiling and/or trade policy play? Issues related to international market stabilisation will arise, since they influence decisions taken by national governments, but emphasis will be placed on domestic policy in light of recent events in international markets, and possible changes in the economic environment within which those policies are set.

This chapter will first explore those changes in the economic environment. It will examine whether world markets are now more volatile, and whether that international market volatility is now more important than domestic sources of volatility (*e.g.* droughts). It will consider the evidence that has been assembled, particularly over the last two years, on the extent to which world prices were transmitted to domestic markets, and if not what blocked that price transmission. Then issues related to the objectives behind stabilisation policy will be explored. That section will consider the disconnect between objectives of international donors and national governments. In light of these objectives, issues related to the choice of policy instruments will be discussed. Particular emphasis will be placed on trade policy and stockpiling. Then the institutional arrangements that condition the effectiveness of policies will be examined, with emphasis on those market institutions (public and private) that affect trade and stocks. The conclusions will summarise policy recommendations and suggest an agenda for future research.

4.2. Economic environment

The World Bank (CRMG, 2008) has asserted that there is considerable unhedged risk in developing country agriculture, and many factors beyond world price variability contribute to that risk. Anton (2009) has argued that the multiple sources of risk in agriculture need to be addressed in a holistic fashion, taking into account interactions among risk sources. Most developing countries have a long way to go to follow those recommendations, but are most concerned with both domestic production variability and variability in the costs of imports to meet food needs. They have pursued partial strategies focusing on price stabilisation, which is in part a consequence of the state of institutions - like commercialisation of agriculture and financial markets. In the absence of the kinds of market institutions found in OECD countries, price stabilisation has been pursued both to balance consumer and producer interests and to mitigate poverty and hunger. While in the longer run it is clearly beneficial to have private market institutions develop, and policies are needed to facilitate that, for now the concerns that drove countries to stabilise domestic market prices persist and may even be stronger.

The price dynamics of 2007-08 have led some to assert that international commodity markets are now more volatile and the tradeoffs between trade policy alternatives and domestic measures including stockpiling are different, with concern over international price instability once again mattering (von Braun, 2008; Delgado and Townsend, 2009). Three issues related to domestic versus international sources of risk to agriculture and to food security are examined in this section: Has world price volatility increased and is the distribution of world prices now somehow different? Does world price variability now matter more than domestic inter-seasonal and inter-annual price variability? Has the transmission of world price variability to domestic markets changed?

World price volatility

Figure 4.1 included two graphs of world grain prices over time, encompassing different periods (1998 to 2009 and 1960 to 2009) to highlight the fact that one's time perspective is critical to assessment of world price volatility. If one has a short memory limited to the previous decade, the events of 2007-08 stand out as exceptional. But a longer perspective shows that world grain prices have exhibited both long periods of stability, like 1998 to 2006, and periods of substantial volatility more like the 2007-08 period. Had we deflated world prices, the variability and peaks of prices in the 1970s would appear more volatile even than recent years. It was recognised in the work on stocks in the 1970s and 1980s that prices were unlikely to follow normal distributions, and that distributions characterised by long periods of low stable prices and brief periods of high prices make stabilisation via public stockpiling strategies costly. This is because large production shortfalls may be infrequent, and distributions may be asymmetric. Moreover, private stockpiling mitigates effects of production shortfalls except in circumstances where prior production or policy has reduced stocks to low levels, contributing to the dynamic of short peaks followed by long periods of low prices. The observed result has been large stocks held over several years, and the rare event that production is low when stocks are also low.

Coefficients of variation are the measure most commonly used to assess price volatility (*i.e.* Balcombe, 2009; Gilbert, 2009; Hazell, Shileds and Shields, 2005). As mean prices increase an increase in variances of those prices should not be surprising, as would occur with neutral inflation. An increase in a coefficient of variation is therefore a more meaningful measure of increased volatility. Anton (2009) used a very similar concept of volatility, based on the per cent change in prices between consecutive periods, to show that by this standard there was a dramatic increase in international grain price volatility in 2007 and especially in early 2008. Figure 4.2 shows his volatility measure for Chicago Board of Trade (CBOT) nearby wheat futures prices as well as weekly SRW Gulf prices alongside the SRW Gulf wheat price. From 1980 until 2006 his volatility measure hovered between 20% and 30% of mean price, and jumped to nearly 60% in early 2008 for daily CBOT prices. The FAO (2008a) similarly noted this rise in volatility of international and US prices, especially in early 2008, for rice, maize and soybeans as well as wheat. Evidence is less compelling for prices after the mid-2008 peaks, relative to a longer history of price behaviour.

Causes of volatility and high world prices

While observations of price behaviour led some to simply examine price variability, others have sought to relate both spikes and variability to causal factors behind high prices. The debate over the causes of price increases in 2007-08 remains controversial, with some areas of agreement. Abbott, Hurt and Tyner (2008, 2009) argued that three sets of factors lay behind high food prices. Supply and utilisation trends resulting in low stocks gave rise to conditions where supply shocks (*e.g.* droughts) mattered more than in earlier years. Exchange rate depreciation and possibly speculation led to financial pressures raising prices, especially in dollar terms. Price increases have been more than the proportional changes in response to exchange rate fluctuations that are suggested by the law of one price, suggesting overshooting may be occurring. Emergence of the biofuels industry and US biofuels mandates linked oil prices and agricultural prices more tightly than before, so oil price increases were passed to grain and oilseed prices – at least for a time. Gilbert (2009) has argued that common factors such as exchange rates, financial

Figure 4.2. **Annualised price volatility and cash prices of wheat**

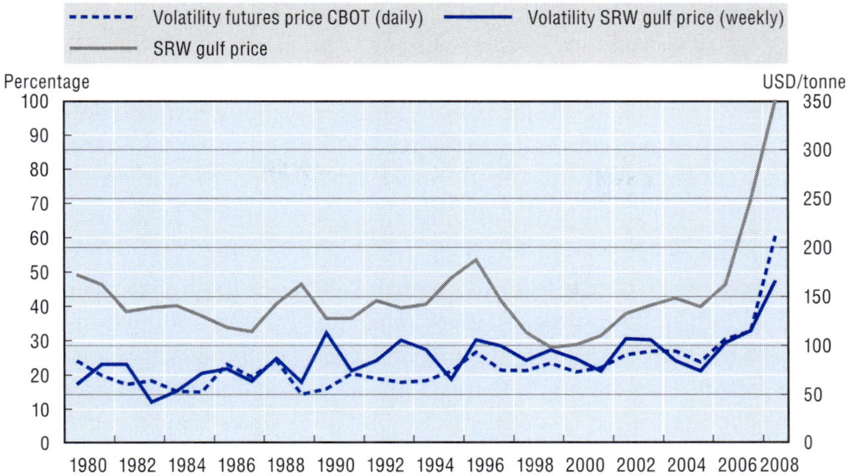

Source: Anton, OECD, 2009, using data from the International Grain Council and Chicago Board of Trade.

StatLink http://dx.doi.org/10.1787/888932552703

speculation and monetary policy must be more important than supply shocks given the coincidence of price increases across commodities. Mitchell (2008), on the other hand, attributed much of the price increase to biofuels, which most directly affects corn and oilseeds.

Debate on the role of speculation remains controversial. Von Braun and Torero (2009) as well as Gilbert (2009) argue for its importance, while Irwin *et al.* (2009) and Wright (2009aandb) believe this factor played a minor role. Some analysts (*e.g.* Timmer, 2008aandb) have included hoarding by domestic agents such as farmers, traders and consumers as part of the "speculative behaviour" that led to the world price spikes of 2008. For example, Timmer argued that rice exports fell in Viet Nam as a result of an export ban, but domestic prices spiked nevertheless as those domestic agents held onto rice in anticipation of future high prices and the lifting of the ban. In assessing the role of speculation in the food crisis a distinction needs to be made between this type of speculation - strategic behaviour of agents engaged in the physical market - versus behaviour of financial "speculators" on futures markets who may not hold positions or have commercial interest in physical markets. While anecdotal evidence of both types of behaviour has related them to price spikes, their relative importance remains controversial.

Most analysts agree that isolationist polices pursued by many countries made world price peaks and market instability greater than they would otherwise have been. Thus, stabilisation policy pursued by national governments not only responded to, but also contributed to the world price spikes that occurred.

Conditional variance of world prices

Balcombe (2009) has attempted to incorporate this information on causes of world price spikes into his assessment of whether or not volatility has increased for many agricultural commodities. He used time series econometrics on monthly data to sort out measures of price variability conditioned by the factors identified as contributing to the 2007-08 food crisis. He found that price variability for grains depends on exchange rates, oil prices and stocks. Once these factors are taken into account, variability is not significantly higher now than in earlier periods, especially if one looks over a longer time horizon.

A lesson to be drawn from Balcombe's (2009) results is that volatility is conditional on market circumstances, and that the distributions of prices need to be understood allowing for differing levels of volatility, driven by those market factors. Uncertainty in markets has increased due to the uncertainty of these factors linked to grain prices, and that these causal factors are themselves very difficult to predict. Wright's (2009a and 2009b) assessment of the role of stocks in determining price volatility is a simple illustration of one factor. His point is that when stocks are large they elastically adjust to quantity shocks, reducing volatility. When stocks are low adjustment becomes more inelastic and prices are more volatile. Hence, the perceived low world stocks in 2007-08 would have contributed not only to high prices but also to higher volatility. But this seems to have occurred at higher stocks-to-use ratios than in the past. Events like the biofuels mandates may also bring structural changes in the factors driving international grain prices. Inspection of oil prices versus grain prices over the last several decades reveals little correlation during the long period of low oil prices starting in the late 1980s, and delayed impacts on the level and volatility of corn prices that became stronger once ethanol plants were online in 2006.

A related question is whether high prices and high volatility occur together, taking into account the expectation that differences in means result in differing variances, but not necessarily differences in coefficients of variation. This depends on what caused high prices. If prices rise because increased costs push output prices – for example, high crude oil prices leading to high fertiliser prices result in high grain and oilseed prices – then it is possible that volatility need not be higher, and depends on the volatility of those input prices. But agricultural prices and hence input costs may be driven by demand, so some have dismissed this argument as an important cause of the high prices in 2008. (Abbott, Hurt and Tyner, 2008). If increased demand (*e.g.* due to biofuels demand) necessitates higher prices as an incentive to greater production, again volatility need not be higher. But if high prices coincide with low stocks, it is likely that volatility is higher with high prices.

Stocks not only signal price levels, but are also the shock absorber reducing price volatility (Wright, 2009a and 2009b).

Domestic versus international volatility

Policy makers who are considering trade policy alternatives to stabilise domestic markets need to be aware of these nuances in the distribution of international grain prices. In 2005, the World Bank observed that domestic price variability was largely due to domestic factors (Byerlee, Jayne and Myers, 2005). Trade liberalisation was recommended as an alternative when variability in world market prices was at a relatively low level. Policy needs to take into account the infrequent but potentially large changes in world prices, which occurred not only in 2007-08 but also in the 1970s and to a lesser extent in the mid-1990s. They need to recognise that there will be episodes of high, volatile prices and of low, stable prices, driven by external factors. They should pay attention to related markets that have caused price increases in the past, aware that new mechanisms arise. They should expect infrequent but large spikes when relying on international markets to smooth domestic markets. It is likely that we may return to a period of stability like 1998 to 2005, where domestic factors dominate, but that spikes in world grain prices can reoccur. Macroeconomic, financial market and energy market factors will influence this distribution. Stocks dynamics, both domestic and international, will also be important.

Domestic variability

While following best practices, and liberalising trade, may have been a problematic strategy during the recent food crisis, the notion that domestic sources of risk and volatility dominate is likely to remain correct in most years. Hazell, Shields and Shields (2005) used several methods and data sources to show the relative importance of domestic production variability versus international price variability. They report coefficients of variation for world rice, wheat and maize prices as well as similar measures for domestic prices in various wheat and maize producing developing countries. In the cases of 18 maize producers, three showed much higher variability in domestic prices relative to world prices for 1971 to 2002, while seven showed lower variability. In the cases of seven wheat producers, two showed much larger variability while five showed much smaller variability over that same period. Differences in domestic volatility depend not only on domestic versus international factors, but also on the extent to which countries stabilise and/or are integrated into world markets. There are several cases with high domestic production volatility yet price volatility less than international price volatility.

Hazell, Shields and Shields (2005) also attempt to distinguish contributions to price volatility due to domestic production variations versus international price variability. Table 4.2 shows that the variance of border prices contributes more than 10% of domestic producer price variance in only 4 of 18 cases, and more than 25% in only two cases. Moreover, there is little evidence that variances in border prices contributed more importantly to domestic price variances after trade reforms. Their data overlap periods when world price dynamics and domestic policy regimes may have changed, and much of the variation in domestic prices remains unexplained in their analysis. But their evidence strongly supports the notion that domestic factors matter more than international prices in most years. Evidence to be presented later on price transmission, used to measure the relationship between world and domestic prices, will tell a similar story, except that persistently high world prices can get transmitted to domestic prices with a lag. While it would be useful to broaden country coverage and bring this research up to date, the relative stability of world grain prices until 2007 means that the conclusions drawn in this earlier study likely remain valid until the 2007-08 period. Which

Table 4.2. **Disaggregation of variance components in producer prices for maize, selected African countries (%)**

Country	Period of analysis	Variance of border prices	Variance of domestic production	Covariance and unexplained	Total
Botswana	1974-1995	1	49	50	100
Ethiopia	1971-2002	1	22	77	100
		(2)	(34)	(64)	100
Malawi	1980-2002	8	1	9	100
		(43)	(4)	(53)	100
Mali	1988-2002	2	28	70	
		(1)	(32)	(67)	
Mozambique	1985-2002	25	2	73	100
		(2)	(1)	(97)	100
Namibia	1980-2002	3	11	86	100
		(1)	(1)	(98)	100
Niger	1971-1995	5	57	38	100
Nigeria	1971-2002	12	26	62	100
		(7)	(16)	(77)	100
Rwanda	1971-2002	5	1	6	100
		(10)	(4)	(86)	100
Tanzania	1971-1995	19	15	66	100
Uganda	1980-1995	1	10	89	100
Zambia	1985-1995	37	5	58	100

Note: Figures in parentheses are for post-reform periods beginning 1991.
Source: Hazell, Shields and Shields, 2005.

characterisation holds in the future depends on our new understanding of world price distributions, considered above, and a subject meriting further research.

Seasonality

If domestic production variability is usually the more important source of instability in a developing country, then it is necessary to understand the short run dynamics of price and production in order to properly design and evaluate policy. Moreover, in poorly integrated traditional markets it may be the case that inter-seasonal price variability exceeds inter-annual price variability, especially in the absence of government intervention. Parastatal grain marketing boards arose in part to combat these price dynamics. The continuing pervasive extent of government intervention in grains markets of developing countries means that extreme inter-seasonal price variability is seldom observed in practice. It is more likely to occur in isolated markets, where data are not collected, as well. There is considerable variability, however, in the extent to which countries succeed at stabilising their domestic markets across seasons.

The extent of inter-seasonal price variability depends crucially on domestic storage markets, which can be quite imperfect. Consider the case of an annual crop, such as wheat or coarse grains, for which production typically occurs once a year, while consumption occurs regularly over the entire year. (Rice would need to consider multiple crops per year.) Stocks must exist to smooth consumption over time, and it is never the case, even in OECD counties, that countries deplete stocks just at the moment the next harvest arrives. Carry-over stocks into the next crop year are often used to gauge supply-utilisation balance and hence inform market price determination. A non-linear relationship is believed to exist between prices and expected carry-out stocks, with lower stocks yielding higher price increases in response to a given quantity shock, characteristic of more inelastic demand when stocks are low (Williams and Wright, 1991; Caifiero *et al.*, 2010). Prices are typically lowest just after harvest, and then increase until the next harvest, when they fall again. This movement of prices over time creates incentives to store. Thus, the extent of price variation over time depends on storage costs, and hence the efficiency of the domestic storage system. Traditional systems might be characterised by mostly on-farm storage in primitive conditions, with high losses and incentives to smooth the consumption of the farm family, but not as much for urban markets. The more commercialised is the market, the greater is the role of marketed surplus, and so too is the role of commercial storage.

Parastatals and storage

When governments replaced traders with parastatals, they also managed stocks. In both developed and developing countries, varying roles have been played by public stocks versus private commercial stocks to smooth consumption and inter-seasonal price variability. One consequence of privatisation of grain parastatals under structural adjustment programs is greater reliance on private storage, but weak institutions and market imperfections may limit how well those private storage systems function after reform.

Both public and private stocks will follow the inter-seasonal price dynamics conditioned by expectations on upcoming harvests and on storage costs. The better functioning are private stocks, the less is the variability of domestic prices, and the easier it would be for public interventions to stabilise prices. Better domestic private storage means lower transaction costs, and probably smaller post-harvest losses. Policy needs to specify rules to manage stocks and trade together, to reduce inter-seasonal price spikes, and to rebuild stocks after bad years. But inconsistent signals from government interventions can disrupt the functioning of private storage markets and timely delivery of supplies and imports (Jayne and Tschirley, 2009).

Stocks and trade over the short run

Both the peak price before the next harvest is realised and the price after harvest are informed by expectations on upcoming production. If the next harvest is expected to be poor, peak prices can be very high as stockholders hold onto grain for the next crop year, and prices after that poor harvest will fall less that in good years. In extremely bad years prices may not fall at all. The period of peak prices before harvest is when food security issues can be most severe, as food is most expensive for those who must purchase to meet their needs. In West Africa there is a French name for this hungry period – *soudur* – and farmers may employ cropping strategies to realise an earlier harvest at higher cost to mitigate this hunger. Private coping strategies in traditional markets can vary from crop management decisions to holding large on-farm stockpiles in anticipation of potential shortfalls (Plateau, 1991). Parastatal market managers would also intervene to reduce these price peaks, when demand could become extremely inelastic so that quantity (or expectations) changes would have larger than normal price implications. Managing inter-seasonal stocks, both on-farm and nationally, is a delicate balance, because it is often the second year of drought that causes the most serious malnutrition, when stocks were run down in the prior year to maintain consumption but are now too low to manage future prices and consumption.

One implication of this short-run perspective is that stocks are an integral part of agricultural markets and annual carry-over stocks always exist, though they can be public or private to varying degrees. Carry-over stocks are the result of inter-seasonal price dynamics, and inter-seasonal stockholding. The more commercialised the market, the greater is the role of private stocks and the easier is the job of inter-seasonal market stabilisation, as that is already being done to some extent by the private market.

A more nuanced view of the role of international trade is called for, as well. If one takes only an annual perspective, it is often argued that imports can make up for production shortfalls and there is no need for carryover stocks, especially public reserves stocks. Only when domestic prices fall between parity bounds – when marketing transaction costs mean it is too expensive for the country to either import or export, so prices fall between these two border prices – would stocks influence domestic prices in that model. Under most circumstances, theory would suggest that, for small trading countries, domestic prices should equal international prices plus or minus border interventions (*e.g.* tariffs). This, and high costs of stockholding, led to the recommendation that imports and trade policy were a better alternative than stockholding strategies to stabilise domestic markets (McIntire, 1981). But imports do not arrive instantaneously, nor necessarily continuously throughout the year. As a season plays out, and expectations change, both domestic prices and plans for later imports vary with these changing expectations. Early warning on the upcoming harvest is crucial to insure that imports arrive in a timely manner to smooth domestic prices. Predicting harvest in advance can be difficult, and this task usually falls to the government, as information is a public good. Poor information and mistakes show up in greater domestic price variability. Stocks interventions may be used to quell price peaks until imports arrive and rebuilding stocks via imports can reduce the consequences of two bad harvests in a row. Trade policy to stabilise prices must take into account inter-seasonal price dynamics, as well as lags in import delivery and the reliability of information on domestic market conditions.

Market imperfections and risk

Newberry and Stiglitz (1981) argued that the justification for price stabilisation policy (or other, possibly better risk management strategies) was due to market imperfections. Missing or incomplete storage, insurance, futures or credit markets are the rationale for government interventions, and the nature of market failure should guide the type of intervention pursued. They argued in 1981 that extensive imperfections existed in developing country grain markets, and recent surveys still share that assessment (Byerlee, Jayne and Myers, 2005; Anton, 2009; CRMG, 2008). The extent and nature of these market imperfections varies by country and probably imperfectly with the stage of development, depending on the extent of commercialisation of agricultural markets. Investments in infrastructure (roads) and institutions (*e.g.* market information systems and legal frameworks) condition the extent of market failures. Traditional and commercial agricultural markets differ in a number of dimensions, and storage is clearly one of them. Traditional markets may have many small traders, and high transaction costs. As markets evolve large traders exploit scale economies but may have market power. They also bring better access to credit, which is crucial to effective risk management, and to commercial storage options. According to Anton (2009), imperfections persist even in developed countries, influencing the design and effectiveness of both public and private risk management tools. While institutions evolve to offer insurance and forward pricing options in developed country markets, stabilisation issues in public policy design persist.

Risk layers and market failure

In his conceptual framework for risk management, Anton (2009) identifies three layers of risk faced by farmers (and countries). The risk retention layer occurs in the middle of the distribution and is managed by farm or household strategies. In both developed and developing countries farm households and consumers will have cropping and saving strategies to cope with these events. In the developed country context, the insurance layer is further out in the distribution and can be addressed by private market instruments such as crop insurance or forward pricing. Anton identifies the extreme tails of the distribution as the market failure layer, where intervention is required. If one thinks about adapting this to developing country circumstances, where countries set policy to intervene and private market institutions are immature, the best practice recommendation would be to adopt a liberal trade policy, with the insurance layer being used to smooth domestic production variability. The market failure layer is deeper in developing countries, and dependent on the state of institutional development. If the country lacks risk management institutions, the insurance layer may be part of the market failure layer. One might think of the food crisis of 2007-08 as an event in the market failure layer, where extremes in the international market were encountered and trade could no longer provide the stability desired by an importing country. By that analogy, interventions of governments to stabilise, and to change trade policy in the face of the infrequent spikes in world prices, is a strategy to address such market failure. Parastatal grain board managers have described their task in this context, in which they need to adopt strategies to cope with catastrophic events (like 2008), but will allow smaller variations in market prices.

Price transmission

Price transmission has been studied extensively to explain developing country response and impacts of high international food prices in 2007-08 (*e.g.* Dawe, 2009; Daviron *et al.*, 2009; Torrero, 2009; WFP, 2009). Studies have examined both the effect of world price changes on domestic prices, and on the extent of domestic market integration – including between urban and rural areas.

Price transmission elasticity estimates in a simplified, ideal world equal unity. Hence, prices in integrated markets would be expected to move under a proportional relationship. Transactions costs, policy interventions, and reversals or cessation of trade flows can complicate measurement of price transmission, and lead to estimated price transmission elasticities less than one. Nevertheless, if one takes trade flows and transaction costs into account, these parameters can inform the extent of both domestic and international market integration (Brooks and Melyukhina, 2005; Balcombe, Bailey and Brooks, 2007).

Border price transmission

Border price transmission estimates the extent to which changes in international prices result in changes in domestic prices, or how well integrated a country is into international markets. According to the annual modelling perspective, differences between border and domestic prices should depend on trade policies (tariffs) and transaction costs. While trade policies often vary to stabilise domestic markets, and were used to this end in the face of the 2007-08 food crisis (Demeke *et al.*, 2009; Abbott, 2009), less endogenous variability is expected in transaction costs. But landlocked countries with high transaction costs would be found by these measures to reflect imperfect integration into world markets, and apparent failure of the "law of one price". To the extent that imperfect price transmission reflects poor domestic world market integration, it is capturing one of the important market imperfections relevant to stabilisation policy and to the role trade policy may play in stabilisation. It also highlights the persistent role played by governments in domestic grain markets.

Estimation of price transmission elasticities is complicated by the dynamics that have been observed in the relationships between domestic and world prices (Baffes and Gardner, 2003). Modern time series econometric methods demand long series, over which price transmission relationships may not remain stable. Many of the studies of the 2007-08 event relied on simple methods due to the very short duration of data series available. Lags were evident in the response of domestic prices to the world price shocks in most cases (Abbott and Borot di Battisti, 2009). While it is expected that it may be eventually too costly for a country to resist changes in world prices for an extended period, in the short run domestic prices may move more slowly than world prices, or may be volatile but independent of world prices. Short-run lagged adjustment dynamics may be the result of either government interventions, poor market integration, or both.

Evidence on border price transmission during the food crisis

Parastatal grain marketing boards were notorious for implementing pan-territorial, pan-seasonal pricing strategies, creating stable prices for farmers (to the extent that the policy regime was stable). These pricing policies were heavily criticised for ignoring transportation costs and incentives to store. While privatisation eliminated these institutions for many countries, it often did not eliminate the desire for stability, so government interventions to stabilise markets continued or returned. Price transmission evidence from several countries exhibits a number of different cases, depending on the circumstances in that country. Three regimes are used in Figures 4.3-4.6 to highlight differing outcomes not only across countries, but also between crops within a country (depending on tradability of the commodity). Figure 4.3 illustrates effective price stabilising regimes in China and Morocco. Figures 4.4 and 4.5 show the consequences of tradability for Burkina Faso and Mali, contrasting rice with sorghum. Figure 4.6 demonstrates consequences of volatile domestic markets, poorly integrated with world markets, for Malawi and Ethiopia. In each graph domestic prices are compared to world prices measured in both dollars and in domestic currency, as exchange rate variations also influence border price pass-through.

Price patterns in China reflect the stabilising policies typical of many Asian countries (Timmer, 2008; Cummings and Gulati, 2009). Figure 4.3 shows that Chinese domestic prices were essentially unaffected by world market events. The small rise in Chinese domestic rice prices is more likely due to inflation than to the spike in world prices. The estimated price transmission elasticity is only 0.15 over the period when world prices rose dramatically. Other countries have also sought to stabilise domestic markets, with varying degrees of success that depends in part on the credibility and effectiveness of government policy. Morocco is an example of a successful North African country in this respect, as shown in Figure 4.3. Morocco has for a relatively long period maintained a very stable domestic wheat price in spite of institutional change driven by structural adjustment reform. Morocco faces high domestic production variability and imports a substantial share of its food needs. Its government is committed to a stable domestic market, however (Tyner, Serghini, and Ouraich, 2010). In the face of the 2007-08 food crisis, it began by cutting high wheat tariffs and resorted to subsidies when

Figure 4.3. **Rice and wheat prices in stabilising regimes – China and Morocco**

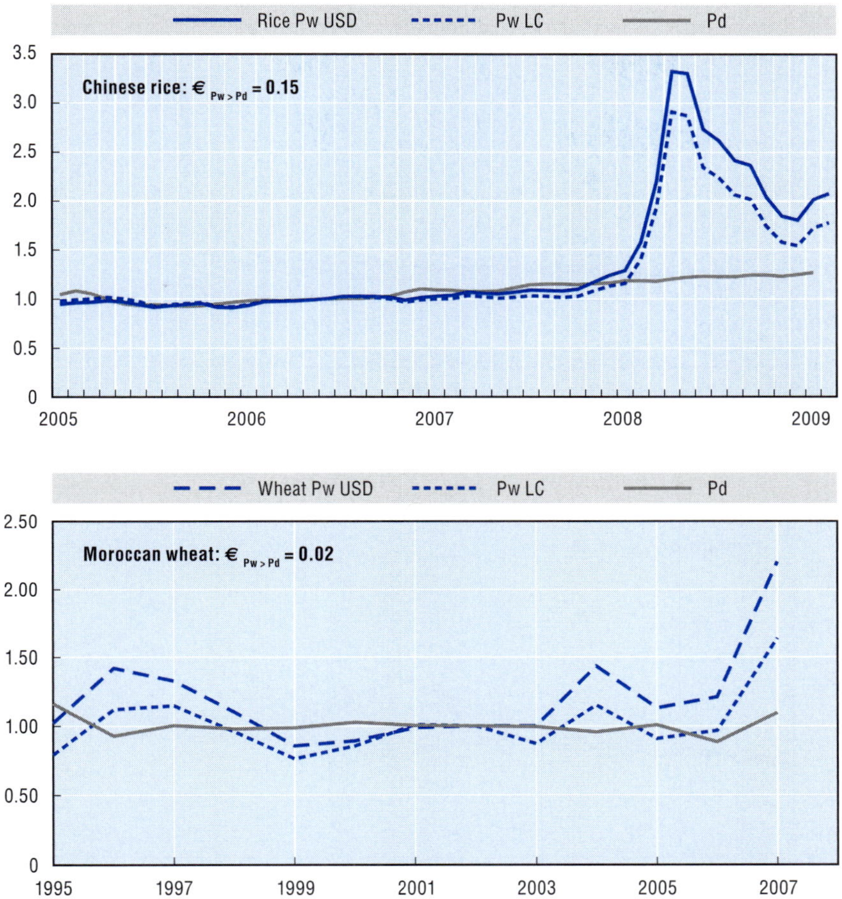

Source: Indices calculated from data in IMF, *International Financial Statistics*, 2010; FAO, GIEWS, 2009; and Tyner, Serghini, and Ouraich, 2010.

StatLink http://dx.doi.org/10.1787/888932552722

the tariff had been driven to zero. The estimated annual price transmission elasticity for Morocco is 0.02 from 1995 through 2007.

The cases of Burkina Faso and Mali illustrate price transmission for a non-tradable staple (sorghum) versus an imported grain mostly serving urban consumers (rice). In several African countries, effects of world prices were more strongly felt on rice than on domestic staples (Abbott and Borot de Battisti, 2009), following the patterns in Figures 4.4 and 4.5. Rice prices in Burkina Faso increased over 60% in 2008, but they lagged world price increases. They increased much less than even the local currency border price, and persisted at a higher level as border prices fell, though some margin

Figure 4.4. **Tradable versus non-tradable grain prices in Burkina Faso**

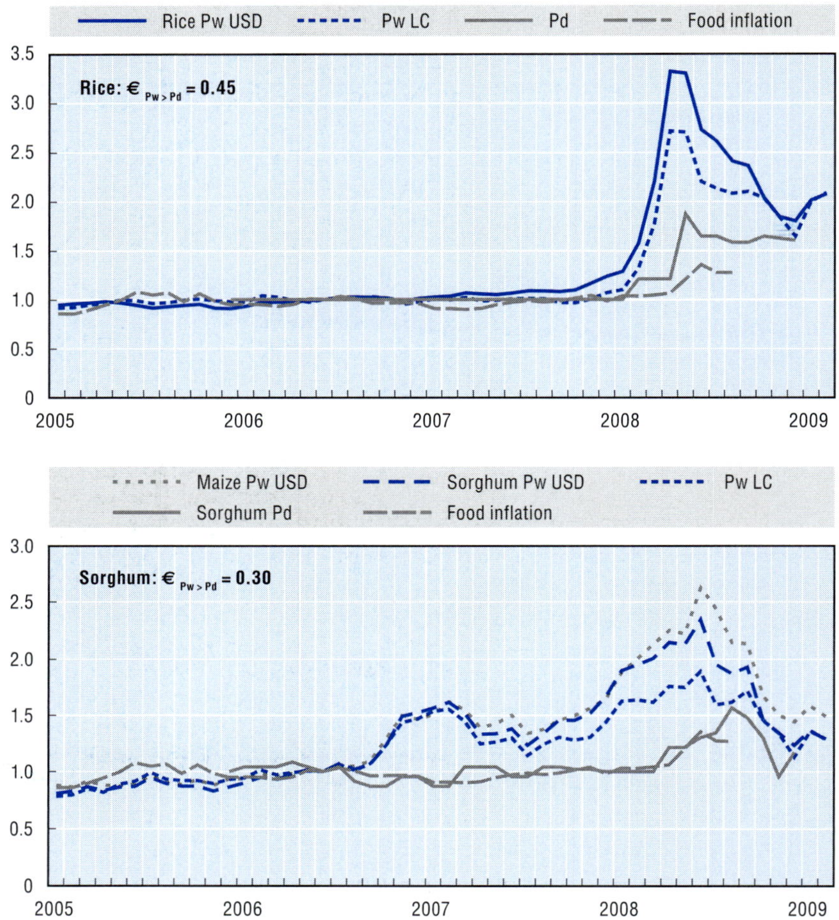

Source: Indices calculated from data in IMF, *International Financial Statistics*; FAO, GIEWS, 2009; and Fulponi, OECD and FAO, 2009 for food inflation collected from various national sources. Adapted from Abbott and Borot de Battisti, 2009.

StatLink http://dx.doi.org/10.1787/888932552741

between border and domestic price changes remained. In the case of sorghum, lags were much longer and it is difficult to sort out pressures from inflation versus border price changes. In the case of rice, the price transmission elasticity was 0.45, whereas it was 0.3 for sorghum. Both tariff and domestic tax reductions were used to stabilise the domestic price in 2007-08, but changes in those instruments were small relative to the world price shock. Mali produces rice and is less dependent on imports, so it saw more stability than Burkina Faso with its rice price increase limited to about 25% and its transmission elasticity only 0.22. Sorghum prices in Mali were more variable than in Burkina Faso, but largely unrelated to world prices, as

Figure 4.5. **Tradable versus non-tradable grain prices in Mali**

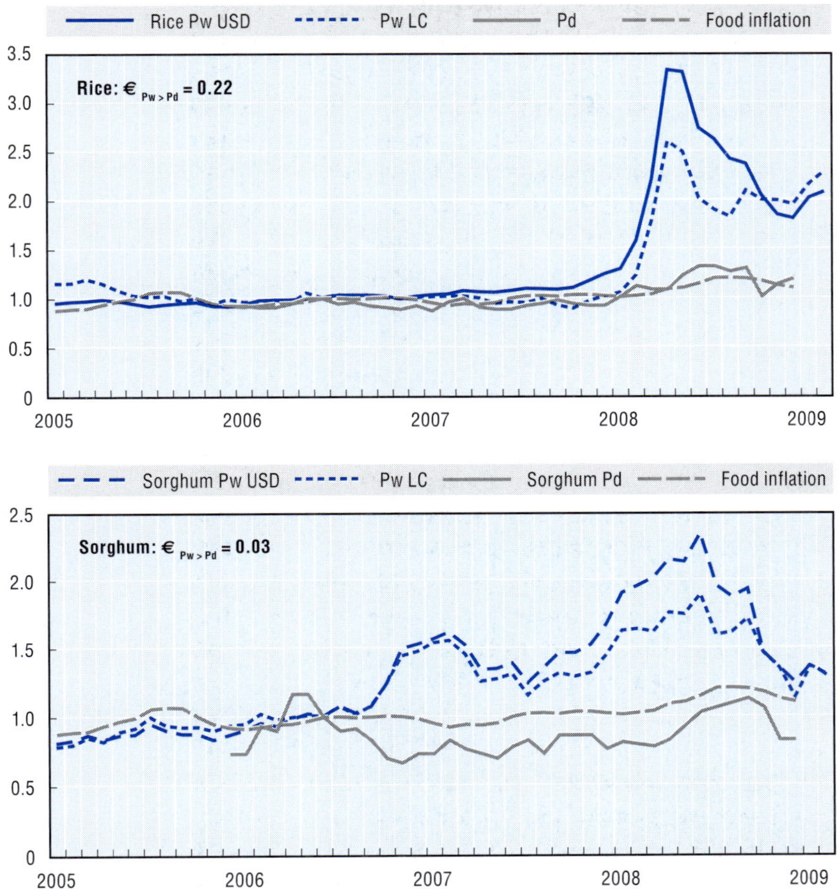

Source: Indices calculated from data in IMF, *International Financial Statistics*; FAO, GIEWS; and Fulponi, OECD for food inflation collected from various national sources. Adapted from Abbott and Borot de Battisti, 2009.

StatLink http://dx.doi.org/10.1787/888932552760

the transmission elasticity was 0.03. In both cases there were long lags between border price changes and domestic price changes.

Figure 4.6 illustrates the cases of Malawi and Ethiopia, where there are highly variable domestic prices, but with changes that do not correspond well with world price changes. There is considerable controversy in the literature on what is driving Ethiopian food markets (Negassa and Jayne, 1997; Loenig, Dureval and Birru, 2009). Before the food crisis, imports were small and were often food aid rather than commercial purchases. Domestic production variability seemed to drive prices before 2007, but domestic price increases in 2007 and 2008 coincided with world price increases in spite of adequate domestic supply or little evidence of more important commercial imports.

Figure 4.6. **Grain prices in volatile domestic markets – Malawi and Ethiopia**

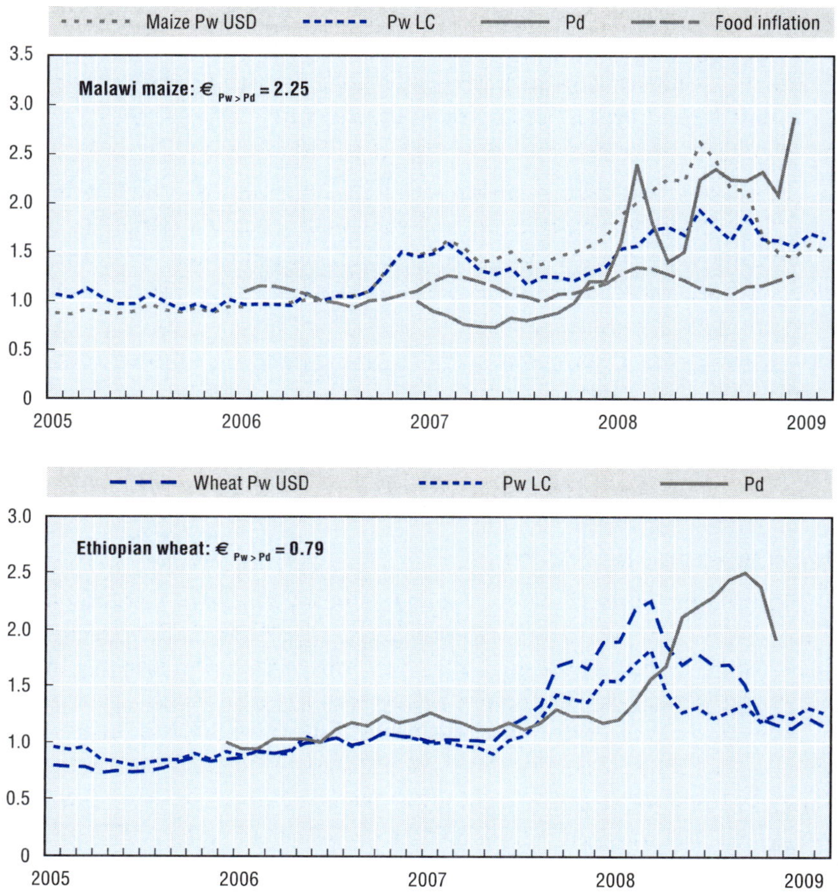

Source: Indices calculated from data in IMF, *International Financial Statistics*; FAO, GIEWS; and Fulponi, OECD for food inflation collected from various national sources. Adapted from Abbott and Borot de Battisti, 2009.

StatLink http://dx.doi.org/10.1787/888932552779

Malawi also exhibits price variability apparently driven largely by domestic factors, but with pressure brought on the domestic market by high world prices during the crisis. Malawi was also influenced by South African maize prices, which did not exhibit the same spikes as US prices (NAMC, 2009). Jayne and Tschirely (2009) argue that mismanaged stabilisation efforts by the government help to explain erratic outcomes in Malawi.

Privatisation and structural adjustment brought not only privatisation of storage markets, but also greater openness to trade. Many countries increased substantially their share of domestic consumption supplied by imports after these reforms. But transmission of border prices to domestic markets often remained incomplete. Both Baffes and Gardner (2003) and Hazell, Shields and Shields (2005) did not find significant increases in transmission of border prices to domestic prices after trade liberalisation. This suggests imperfect market integration persists as policy barriers are removed. The extent of stabilisation evident in some developing country border prices is greater than could be achieved by the border and even domestic policy changes that were used to stabilise domestic markets (Abbott and Borot de Battisti, 2009).

In each of the cases illustrated here, and in many other developing countries, spiking world prices put pressure on both domestic markets and on governments to intervene. While trade policy tools seemed to be of limited effectiveness, price transmission was incomplete and lagged behind world price changes. There is substantial variability among findings on price transmission across both countries and commodities, due to differences in policy effectiveness, government commitment, and the state of domestic market institutions. In general, Asian countries effectively stabilised while African countries were less able to resist world market pressures, but important exceptions exist in each region.

Domestic market integration

Price transmission is used to gauge domestic market integration as well as integration with world markets. Studies of the 2007-08 food crisis looked at pass-through of border prices to both the farm gate and to urban consumer prices. In some instances, border prices increased urban consumer prices much more than farm gate prices (Torrero, 2009 from IFPRI studies of Latin America). In other cases little difference was detected, with border price changes passed to farmers or with urban and rural prices integrated with one another, but not with border prices (Daviron *et al.*, 2009; Dawe, 2008; WFP, 2009). Past studies of domestic market integration have found variable results, depending on the extent of infrastructure, institutional development and policy. Greater commercialisation is likely to lead to better market integration, so greater pass-through, at least of urban prices to rural areas. It is not uncommon even to find differences across regions within a country. More remote regions may be poorly integrated with urban and world markets, but where roads have been built, and markets developed, domestic price transmission is higher. Where price transmission is greater, and market integration is better, production shortfalls are more likely to be made up by supplies in neighbouring regions.

One model which can reconcile some of these findings, particularly the lags in price transmission, is average cost pricing by large private grain traders. That model presumes high-fixed transaction costs and scale economies in marketing and distribution, a likely circumstance. In the face of rising world prices, traders would only slowly raise prices to their customers, and continue to pay low prices to farmers in their own country. Keeping prices paid to farmers low would make this strategy more affordable. Arbitrage opportunities would force world and domestic prices to converge over time, but with few traders in a market this could be a slow process. Parastatals may have employed such a strategy, and are less likely to face new entrants and pressure from arbitrage opportunities than private traders. Whether this is a correct model remains to be tested, but this behaviour can help to account for the degree of stability observed in many developing countries' grain prices following the 2007-08 food crisis, the greater change in urban versus rural prices and lags in price adjustments. Both scale economies in marketing and distribution and market power of traders need to be considered in formulating stabilisation strategies. Trade modellers to date have preferred to model these markets as competitive, but there is some evidence from agricultural export markets supporting the assumptions behind this explanation.

4.3. Objectives

Policy debates often involve interest groups and analysts arguing cross-purposes, and stabilisation policy debates are no exception. Before evaluating policy instruments used to stabilise domestic prices, and alternatives that address risk more broadly, it will be useful to examine the issues and national goals that lead countries to stabilise. Several questions must be addressed: why do countries so often stabilise, should they stabilise, and if so, what should they stabilise?

Table 4.3 illustrates the various objectives that may be taken into account as national governments formulate and implement stabilisation policy. Most economic analysis starts with basic economic welfare – consumer surplus, producer surplus, and government revenue or cost. In order for interventions to be justified, in most cases either some market failure needed to influence these basic outcomes, or a social objective function needed to weigh these criteria differently or address additional social goals. Early work on stabilisation policy focused on basic measures of economic welfare, while more recent literature examines why stabilisation seems to persist in agricultural policy regimes.

Basic economic welfare

Debate on stabilisation policy initially focused on whether producers or consumers would realise higher welfare under a price stabilising regime, relative to when there was no intervention. The question addressed was whether stabilisation could raise basic economic welfare, either of specific agents or of a nation. It was found that agents' welfare depended on the source of shocks (supply or demand), the specification (shape) of supply and

Table 4.3. **Objectives relevant to stabilisation policy choices**

Basic economic welfare		
	Producer surplus – net farm income	
	Consumer surplus – utility	
	Government revenue – and so government cost	
Correcting market failure		
	Distribution infrastructure	• Market information • Commercial storage
	Credit – missing or incomplete markets	
	Risk aversion	• Insurance and risk management – lagging private institutional development • Risk and agricultural production decisions
Social objectives		
	Income redistribution – political economy	• Farmers versus consumers • Special interests matter • Willingness of government to incur costs
	Food security – of urban constituencies	
	Poverty – rural and urban	
	Stability as a preference	• Price stability • Macroeconomic spillovers – agricultural stability implies macro stability
	Avoiding extremes – in distribution tails	

demand curves, hence specification of welfare functions, and the nature and extent of transaction costs. For example, linear supply and demand held different implications for welfare outcomes of producers versus consumers under stability than did a constant elasticity specification. This literature also asked if net social welfare could increase in the absence of distortions, but with price instability.

Newberry and Stiglitz's book (1981) was considered one of the definitive works addressing these issues in a theoretical framework. Anton (2009) and Byerlee, Jayne and Myers (2005) reviewed the extensive literature on welfare implications of stabilisation and examine broader risk management strategies. High transaction costs and asymmetric distributions mean that net welfare gains are unlikely unless stabilisation corrects other distortions, but there may be both winners and losers under a stabilisation regime.

Early in the literature, Newberry and Stiglitz (1981) argued that price was the wrong variable to stabilise, even if it is the most politically expedient choice. Revenue or income is a more appropriate choice for farmers, and from a national perspective consumption or social welfare should be in the objective function. Moreover, Newberry and Stiglitz (1981) argued that price stabilisation can destabilise farm income under the right circumstances. Like Newberry and Stiglitz, much of the literature has examined farm income effects more so than national welfare. But prices are easily observable and have been used to balance competing political interests. Most work subsequently considered the nature and extent of market imperfections, and how price stabilisation or other risk management strategies interacted with these market failures.

Focus on farmers in the design of stabilisation policy arose because agricultural commodity income is a large part of their income and because rural poverty is more prevalent that urban poverty (World Bank, 2008). The recent food crisis, however, probably impacted urban consumers more so than farmers or even the rural consumer because urban price increases were likely not to be fully transmitted to farmers. While most rural residents may be net food buyers, if they were farmers they did not have to buy their entire grain requirement. In 2005, the World Bank (Byerlee, Jayne and Myers, 2005) recognised that consumer welfare is an important component of policy assessment. Actions taken by developing country governments reflect concern with urban consumers broadly and not just the extreme poor or farmers (Abbott, 2009). These decisions show that social objective functions take into account income distribution, so go beyond basic welfare functions.

Addressing market failure

A principal justification for government intervention is to address market failure or missing markets. In the case of domestic grain markets, as noted above, it is likely that in developing countries there are missing or imperfect risk and insurance markets. Moreover, imperfect credit and financial markets, and lagging institutional development affect the conditions for storage and trade. Parastatal marketing boards arose in part to address those and other market failures, including the possibility of market power of domestic traders. Pan-seasonal and pan-territorial pricing addressed weak market integration, so the state not only stabilised, but also addressed other market failures. But it has been argued that even where these public institutions work well, at least at attaining stabilisation goals, state-run markets are an inefficient, costly alternative. As structural adjustment brought private trade in place of parastatals, some markets have run more efficiently, while some of the market failures these entities previously addressed have become apparent. Storage and finance markets demonstrate the need for institutional development and policy interventions, but stability as a consumer or producer objective also persists. Galtier (2009) notes that the private risk management strategies advocated by the World Bank (CRMG, 2008) have not taken hold or addressed the desires for stability in these markets. Others argue that desires for stability led many governments to return to stabilisation after structural adjustment reforms (Byerlee, Jayne and Myers, 2005).

Risk aversion

If agents are risk averse, then reducing the variability or extremes that agents face raises their welfare. Both price stabilisation schemes and private risk management strategies seek to realise such welfare gains, by reallocating risk from those who are risk averse and to speculators who prefer high potential reward over risk reduction. Therefore, to the extent that risk aversion characterises either farmers or consumers, their objective functions value these "risk benefits" of stabilisation. Public intervention is required when private markets, such as insurance and futures' markets, are imperfect or missing. Early on as structural adjustment programs eliminated parastatals, the World Bank advocated private market strategies, focusing on futures market options. Literature now finds futures markets to be more appropriate for traders than small farmers, and numerous problems arising from moral hazard, adverse selection and basis risk plague the design of crop insurance alternatives (CRMG, 2008; Byerlee, Jayne and Myers, 2005).

Another market distortion that stabilisation may address relates to incentives to invest in agriculture and so the rate of growth of agricultural production. Boussard (2004, 2006) has long argued that success of EU agriculture, and the shift from importing to exporting, derived more from stability of prices than from high prices. Several authors have argued that the success of the green revolution in India and elsewhere was supported by the stable price regimes that were in place (Cummings and Gulati, 2009). Risk averse farmers would make decisions in favour of lower yield but more stable crops, would use fewer purchased inputs, would invest less, and would be less likely to commercialise if prices were not stabilised (Timmer, 2002). It may simply be the case that agricultural production under stability grows faster.

Byerlee, Jayne and Myers (2005) argue that estimates of the welfare gains to farmers from stabilisation, done mostly for export crops, are found to amount to only about 2% of the value of agricultural production. While poor countries who depend on a single staple, and who are landlocked or spend a high fraction of foreign exchange earnings on food, may be vulnerable, empirical estimates of the value of these effects have proven difficult. The extent to which countries pursue stabilisation policies means that either they see these benefits as being more valuable than these estimates suggest, or that they operate under a different objective function than is presumed in standard analysis. For example, they may be implicitly pursuing income redistribution objectives.

Social objectives

Another rationale for policy intervention beyond market failure is that social objective functions may differ from the basic welfare measures of economic models, based on private objectives summed. We have already seen risk aversion (which was coupled above with risk market failure) and from a consumer perspective food security may also be a national objective that policy addresses. Political economy models highlight that consumers and producers may be counted differently in social welfare functions. It is not uncommon in developed country agricultural policy models to apply higher political weight to producer welfare than to consumer welfare. Hayami and Anderson (1986) describe a transition from higher weight on consumer welfare to higher weight on producer welfare as economies develop and as agriculture becomes a smaller share of an economy. Consumers have historically counted heavily in developing countries.

Abbott (2009) argued that the differing responses of developing country national governments and the international donor community is explained in part by differing social objective functions, and in particular differing emphasis on the political weights attached to urban consumer welfare versus small farmers and the extreme poor. International donors emphasised safety nets and agricultural production investments, with emphasis on small farmers. Their objective function weighs more heavily both rural residents and the extreme poor. National governments clearly put more weight on urban consumers, and have a higher poverty threshold than is evident from international donor recommendations.

Poverty and food security

Addressing poverty is another dimension to the social welfare function that may lie behind stabilisation policy. The poor may be more vulnerable to risky events, and may have access to fewer strategies to cope with the risks they face. The consequences of high price, high food cost or low price, low income scenarios may also be more severe. If prices are high and diets are inadequate, malnutrition may bring both short and long run costs. At the extremes, survival may be at stake. Malnutrition may also affect labour productivity and an individual's capacity to work. The poor also spend a larger share of their income on food. Thus, food insecurity is an important special aspect of poverty (Rosen and Shapouri, 2009; FAO, 2008b). One goal of price stabilisation schemes is to avoid or reduce malnutrition, famine and even poverty.

Strategies to address food insecurity and poverty often target food consumption of specific groups, as do food aid, food subsidies and food for work programs. Recommendations for safety net strategies generally recommend targeted income subsidies over general food subsidies, leaving flexibility for consumers and more efficiently allocating resources to the poor. The consequences of malnutrition mean that food interventions may be more highly valued than income interventions, and food interventions may be more easily supported from a political perspective. If poverty is the main concern, however, targeted programmes are likely to be more efficient than stabilisation policies.

Avoiding extremes

Both the poverty dimension of national government objectives and the asymmetric nature of grain production and price distributions mean that social goals may emphasise avoiding the tails of food price distributions. Moreover, governments are likely less concerned with the small risks associated with small variations in price around its mean and are more concerned with the consequences of extreme variations. Parastatal grain managers often saw as their goal avoiding extremes and catastrophes (the tails of distributions, or the market failure layer in Anton's 2009 framework), rather than reducing variances of prices.

National polices often reflect this nature of their social welfare function by choosing price bands policies – that is, choosing to keep prices above a floor price, below a ceiling price, or within a range. While the nature of their objective function leads to this policy choice, economists have been critical due to potential, theoretical outcomes under such a regime. Wright (2008aandb) argues that when price bands are in place, there are incentives for speculators and traders to store in a way that causes prices to stay near the policy set price bands more so than a distribution without intervention would yield. Salant (1983) went further, arguing that price bands and complementary storage policies invite speculative attacks. Their arguments presume speculators with financial resources can intervene, and transaction costs to do so are low. These conditions are more likely to apply in developed country and international agricultural markets. One of the few advantages of poorly developed storage markets is that the likelihood of such speculative attacks is very low, so governments may be able to utilise price bands with fewer problems from the private market.

Price bands also influence expectations for producers, for better or worse. If producers are risk averse, price floors may elicit greater production growth than would otherwise occur. Again following Anton's framework, the strongest argument for government intervention is in the market failure risk layer, also supporting policies aimed at avoiding catastrophic market extremes. The challenge is for governments to intervene to avoid extremes in a transparent and consistent manner, cognizant of the incentives the regime creates.

Cost

Government cost is also a component of the relevant objective function, and evidently gets a high weight from some developing country governments (and even higher weight from many economic analysts). Cummings and Gulati (2009) argue that the flaw of Asian parastatal grain market management is its high cost, but it is evidently a cost many Asian governments choose to bear. On the other hand, as many governments pursued costly tariff and tax reduction strategies to stabilise domestic grain markets in 2008, they also asked for "fiscal space" in the form of loans from the World Bank to make up lost revenue. This fiscal space alternative accounted for a substantial share of the expenditure by the World Bank in its GSRP programme, its response to the 2007-08 food crisis (World Bank, 2008a). Governments that stabilise have historically continued to import to meet food needs, in spite of the costs due to high world prices. Import demand functions are often very inelastic during crises, as a result. Thus, the importance of costs probably varies widely across national governments, and depends on either the ability of the government to pay high costs, or its ability to attract foreign aid to cover some of those costs. Structural adjustment reforms were often the product of inability of a country to finance these costs rather than buy-in to a more market oriented grain market regime.

Macroeconomic spillovers

Timmer (2002) in particular has also emphasised that there may be macroeconomic spillovers from stability, particularly in countries where agriculture remains a large share of gross domestic product and employment, or grains are a wage good. He uses Indonesia as an example of a country that gained a greater degree of macroeconomic stability, and avoided some of the losses from recession, by stabilising its large and important agricultural sector. Food price shocks can contribute to business cycles, and food price stability can influence investment incentives and so economic growth (Dawe and Timmer, 2005). But cross-country studies have not found a systematic relationship between commodity price fluctuations and economic growth (Macbean, 1966; Deaton, 1992; Kannapiran, 2000).

Policy objectives vary widely across countries, and between national governments and international perspectives. Any policy recommendations will need to reflect the objective function chosen. For purposes of subsequent analysis here, we will take the perspective of a national government in a poor developing country, where arguments for stabilisation are strongest, and which include both urban and rural interests. In spite of the potential (theoretical) problems, policies that intervene at extremes are probably a better reflection of the objectives of those governments. Market failure is sufficiently likely to justify some effort by governments to protect against those extremes. The 2007-08 food crisis would clearly be the case of such an extreme event and of failing markets.

4.4. Policy instruments

Best practices for risk management and stabilisation policy outlined by the World Bank (Byerlee, Jayne and Myers, 2005) emphasise long-run investments that allow the private sector to better cope with meeting food needs of the population. Their second element is to promote diversification and investment strategies such as irrigation, drought tolerant varieties, and cropping-mix strategies, in order to reduce output volatility while increasing production. This comes after a recommendation to improve overall productivity, their first recommendation. Strong emphasis is placed on long-run food availability rather than coping with crises at hand. Their next recommendations addresses institutional development of private food markets, including infrastructure, market information, regulation and co-ordination, and specific private market-based risk measures, such as forward pricing and crop insurance, intended to make markets work to better share instabilities. They state that "... direct public interventions in food markets to manage price risk should be a last resort" (Byerlee, Jayne and Myers, 2005, p. xiv). They want resources reallocated from a short-run "fire fighting" focus to a long-run, private sector led approach. The only short-run measures advocated are safety nets to address poverty concerns in the face of price spikes, preferring cash transfers to food aid. In their view, safety nets should promote private sector development rather than distort markets.

Their recommendations are largely based on the observation that domestic shocks dominate. They argue that trade liberalisation serves as a stabilising force, since world market volatility is less than domestic sourced volatility. The events of 2007 and 2008 seriously challenge this underlying perspective. National governments in the face of these events rejected the best practices advice and pursued expensive short-term measures to mitigate the effects on domestic markets of world price shocks (Abbott, 2009). Earlier analysis of the distributions of world grain prices and domestic production was intended to argue that in most years, the best practice approach of focusing on domestic sources of volatility remains the best strategy, but that governments need to be prepared to address the real, if infrequent world price increases that will occur. Safety nets are important since the poor are most vulnerable, but governments face concerns of a much broader constituency, including urban consumers not among the extreme poor. The question then becomes how (or whether) to use trade policy or stocks, the short-run policy instruments historically used to stabilise, to address those infrequent world price shocks.

Stocks

Two types of stocks are maintained by governments, serving different purposes. Food security stocks or *emergency reserves* are best thought of as working or "pipeline" stocks serving safety net programs. Food distribution schemes, food for work programmes or other food aid initiatives require stocks be held to insure timely delivery of food in those programs. *Strategic reserves* are used to stabilise the market price of a commodity. Stabilising stocks may be publicly-owned and managed, or may be privately held commercial stocks governed by rules set by national governments. Historically in developed countries it has not been uncommon to find both public and private stocks strategies combined to stabilise markets. The rules are probably more important than who owns the stocks, but if the government is to intervene sporadically, it is probably best to base any strategy on privately held stocks, that will exist in any case. Consistent and transparent rules, and institutional improvements to insure well-functioning private storage markets are prerequisite to using a private stocks strategy.

Strategic reserves or trade?

Analysis of strategic reserves is generally done on an annual basis and under the assumption that a developing country is a small country in international grain markets (*i.e.* a price taker). That analysis begins with the supply-utilisation accounting identity relating production, trade, use and stocks:

$$S_t + Q_t + M_t = C_t + S_{t+1}$$

where S_t is carry-in stocks in year t, Q_t is production, M_t is net imports (negative if exports), C_t is consumption (use), and S_{t+1} is carry-out stocks in year t, hence carry-in stocks in year t+1. Several insights follow from this identity, which must hold. Carry-in stocks are predetermined and production is stochastic and determined ahead of consumption. Consumption can be very inelastic, so carry-out stocks are typically the most elastic adjustment mechanism clearing the market, but stocks adjustments become less elastic as those stocks become low. Trade and/or stocks can clear the market each year, and if the small country assumption is appropriate, world price determines domestic price while trade (M_t) clears the market. If this is the right paradigm manipulation of stocks only impacts the trade level and not the domestic price – so stocks are ineffective as a tool for domestic price stabilisation under this perspective.

If world and domestic prices are not linked as would be the case if the country were a price-taker, because world and domestic markets are poorly integrated, transaction costs are prohibitively high, and/or the country's price is between parity bounds, then stocks will clear the domestic market and can be manipulated to adjust domestic price. If a country is a net importer and a price-taker, Pd = Pw + T + cm where Pd is the domestic price, Pw is the world price, T is a specific tariff and cm are transaction costs to bring imports from the world market. If a country is a net exporter, Pd = Pw – ce –Te where ce are transaction costs to bring exports to the world market and Te is an export tax. In a small country case Pw is fixed. Parity bounds mean that the domestic price lies between the price at which it would export and it would import: Pw – ce - Te < Pd < Pw + Tm + cm. In this case there is no trade. In landlocked countries, where cm and ce (transaction costs) are large, hence prohibitive, there can be a wide range of domestic prices that result in no imports or exports ($M_t = 0$). Policy (Tm and Te) can also influence the range of prices within which a country is self-sufficient. All of this assumes the law of one price assumptions hold, and that markets are integrated when the domestic price is outside the parity bounds set by world price and transaction costs.

Within this framework linking Pw and Pd, and if Pw exhibits low volatility, there would be only a few cases where stocks are an effective tool to manage domestic prices. Therefore, under this theory one should observe countries choosing either stocks or trade, not both, to stabilise. If one instrument is effective, the other is not. Trade is likely to be the better choice, unless world and domestic markets are not well integrated.

Observed public stockholding

Examination of supply and utilisation data for developing countries will reveal that different countries adopt different strategies, but that most use a mix of trade and stocks to maintain consumption in the face of production variability. In a study of 18 developing countries in the mid-1990s, Abbott, Patterson and Young (1998) found that stocks adjustments to production shortfalls were statistically significant in 16 cases, making up between 20 to 50% of shortfalls in most cases. Trade adjustment was also statistically significant in all but a couple of cases and typically made up a similar fraction of production shortfalls. Thus, contrary to the dichotomy of the typical trade model, countries have used a combination of stocks and trade policies, with stocks being an important adjuster even when trade occurs.

Earlier evidence on market integration questions the small country assumption in which the law of one price holds, showing at least short run deviations as world prices change. Price transmission is quite incomplete. Moreover, a perspective looking at a shorter time frame and seasonal effects, in which imports arrive only with a lag, and short-term uncertainty on production and hence import needs exists, allows for short run dynamics to set a role for stocks. Lags in import delivery and transmission of world prices to domestic markets influence how far the domestic price can differ from import (or export) parity prices. In markets where domestic production volatility dominates, as was seen for the Ethiopian and Malawian cases, substantial deviations are observed even when there may eventually be international trade. Price deviations depend on how effective are domestic storage markets and on public perceptions as to the extent that government will intervene to stabilise. In cases where government intervention is credible (China, Morocco) even seasonal prices will remain stable.

Short run stocks management

From the short-run perspective, stocks are inevitable. Stocks will never be driven to zero as harvest arrives, and the extent of carry-over will be influenced by expectations on harvest, with prices smoothing consumption across crop years. The extent to which prices are smoothed, or are volatile, depends on domestic storage institutions, expectations on government intervention, and expectations on the extent to which trade will eventually make up shortfalls. As noted earlier, poorly developed marketing and storage institutions can result in greater instability of inter-seasonal prices than of inter-annual prices if governments do not intervene.

The policy challenge in this environment is to manage short-run domestic price dynamics given trade opportunities and given the state of domestic institutions. Rules and incentives will govern the extent to which stocks are carried over in any year, whether they are public or private. The stocks themselves need not be owned publicly, but for the government to control (or at least influence) reserve or strategic stocks via policy rather than ownership, then well-functioning private storage markets are needed. Empirical research has focused on inter-annual rather that inter-seasonal dynamics, so does not inform well how to make tradeoffs in this short-run dynamic context. Policy recommendations must then be based for now on basic principles, bearing in mind this shorter run perspective. One of the "best practice" recommendations that stands up here is that development of market institutions, such as better market information and addressing imperfections in risk markets, surely are needed. Legal reforms, such as warehouse receipts, will facilitate private storage market development. But public good aspects remain for market information, including early warning on crop forecasts in addition to sharing market price information widely.

Pan-seasonal pricing is problematic in this context, in that it suppresses incentives to store. When parastatals stored, those temporal incentives may have mattered little, but as long as private agents also store, distortions to the seasonal pattern of prices will distort stockholding. Even successful stabilisers build in storage incentives to their pricing strategies.

Infrequent price spikes and stocks

The challenge of the 2007-08 food crisis is that a seemingly new distribution of world food prices was revealed, and when the price rose volatility may have increased. The basic assumption about relative volatility of domestic production versus world prices may no longer be valid. A longer historical perspective might suggest this was not a new distribution, but the infrequent spikes that characterise these markets. In principle, countries could hold stocks in anticipation of these spikes, and use those stocks to keep domestic prices from rising in such instances. In practice, those stocks would need to have been held for a very long time, incurring substantial opportunity costs. The stocks literature looking at world markets turned to financial alternatives rather than physical stocks to seek to avoid those costs. From a short run perspective it means each year keeping carry-out stocks at a sufficiently high level to be ready in case world prices spike in the future, and so trade becomes a too costly stabilisation option. Trade may be needed to rebuild stocks in bad years, with the goal to enter each new year in normal rather than shortage status.

In the context of research on international stocks, the search for financial (and trade) alternatives to avoid the opportunity costs, and losses, from holding physical stocks at high levels or for long periods have been considered (Huddleston *et al.*, 1984; von Braun and Torero, 2009). Moreover, institutions have existed in the past at the International Monetary Fund and at the European Union to insure foreign exchange was available so that trade policy alternatives could replace stocks. These programs were little used, as the conditions required to obtain this financing were stringent and high food import costs did not always coincide with low foreign exchange availability.

The high transaction costs of stockholding matter to both strategic reserves and to emergency reserves held to manage safety nets. These costs mean to the extent possible, longer run stabilisation strategies should rely as much as possible on trade rather than stocks. But the feasibility of trade alternatives depends on how well integrated a domestic market is to international markets.

Trade policy

Nearly three decades ago, McIntire (1981) at IFPRI recommended that variable levies rather than stocks be used to stabilise domestic grain markets of developing country importers. If the price linkage defined earlier from the law of one price applies, so that Pd = Pw + T + cm for an importer, in principle tariff changes can counteract any world price changes to keep the domestic price stable. If a country is an exporter, an export tax can play a similar role – and was used to this end frequently in 2008. There are limits to these tools, in the case of imports, when world prices rise and the tariff goes to zero, and in the case of exports when world prices fall and the export tax is zero. Byerlee, Jayne and Myers (2005) had argued against variable levies as a stabilisation tool in part because a very high tariff, hence a very large persistent distortion, was required to protect against possible large world price increases. In each case, a subsidy may then replace the tax when large world price changes are encountered, nullifying their argument. Subsidies to stabilise are costly, but cutting taxes and tariffs also reduce government revenue. The price linkage formula applies in each case, assuming T can be negative in the case of an import subsidy or export tax. In the face of rising prices, tariff reductions bring revenue losses, whereas export tax increases bring revenue gains. As prices increase a nearly self-sufficient country might also switch from being an exporter to being an importer. In that case, reducing tariffs could switch to increasing export taxes.

Cost is always the concern with these border price instrument regimes. The advantage to variable levies over stocks is that holding and transaction costs need not be borne in more normal years, the more frequent case. Thus, trade strategies are expected to be less costly.

Price bands

A popular strategy to accomplish stabilisation was a price bands regime, where governments intervene and vary policy instruments only when floor or ceiling prices are breeched. Chile used this instrument successfully until it was required in the mid-1990s to comply with WTO requirements that these stabilising tools no longer be used. Many of the tariff, tax and subsidy changes in 2007-08 are best viewed in this context as they were not varied until extreme price increases were faced. In the price bands regime interventions only occur when market extremes are faced. As noted earlier, price bands have been criticised, particularly in theoretical literature (Wright, 2009a; Salant, 1983) because these regimes are vulnerable to speculative attacks. Private marketing agents can buy until capacity constraints are met, and then sell when prices eventually rise above the bands. This is more likely to occur in well-organised market with large, well informed traders, so may be less of a concern in poorer countries, and where market imperfections are important. Speculative attacks may also be less likely when price bands are implemented via variable tariffs, as capacity constraints are less likely to bind, though financial constraints could bind. This strategy fits well a national objective that emphasises avoiding catastrophes.

Observed use of trade policy to stabilise prices

The variable levy regime was practiced explicitly and effectively by the European Union to stabilise domestic prices until the 1995 Uruguay Round WTO agreement, and has been used implicitly since to continue to maintain stable domestic prices. In the mid-1990s the European Union also switched to export taxes when world prices went well above domestic price targets. It did not do this in 2008, but did cut its tariff as world grain prices rose. This case demonstrated the power of variable levies to stabilise. It also raised a major concern that countries who stabilise domestic markets via trade export their instability onto world markets (Bale and Lutz, 1979). Trade policy stabilisation is a beggar-thy-neighbour regime. Explicit variable levies are now WTO illegal. The WTO prohibition on variable levies arose because of the consequences of this policy on trade partners. Governments can change "fixed" tariffs, however, as the European Union has shown. Following banning of variable levies by the WTO in the 1995 Uruguay Round Agreement, the European Union has used a "fixed" levy that it changes as often as bi-weekly, based on market conditions. Safeguards as an alternative are controversial, and one of the issues that prevented reaching a new WTO Doha Round agreement. Safeguards also address low prices and import surges, but not high prices. WTO commitments were seldom a concern as policy responses were implemented in 2007-08 (Abbott, 2009).

Both tariff reductions and export taxes were used frequently by developing countries to stabilise domestic prices in the face of the rising world prices (Demeke *et al.*, 2008). Two principal concerns of grain importers during the 2007-08 food crisis were reduced revenue (loss of "fiscal space") and that tariff changes could not be large enough to offset the large world price increases. A few countries resorted to food subsidies to counteract the very high world prices, at considerable expense. Some traditional importers also introduced export taxes or bans to prevent supplies from leaving domestic markets. Domestic policies can also have trade effects, so reductions of domestic taxes or domestic subsidies could be used like trade policies to control prices. Demeke *et al.* (2008) observed that such domestic policy changes were also frequently used in 2007-08 by developing countries. These taxes and subsidies were also small relative to the world price increases that occurred, limiting their effectiveness, and domestic measures were applied to a larger quantity (all consumption, not just imports) so are more costly in terms of revenue loss than are tariffs. Table 4.1 showed the variety of instruments used in 81 developing surveyed by Demeke *et al.* (2008) and the frequency of use of these alternative instruments.

Quantitative trade controls by parastatals

Prior to the 1995 URAA, or until their structural adjustment reforms, many developing countries utilised quantitative controls administered by parastatal marketing boards to effect a similar outcome. These institutions persist, particularly in Asia where structural adjustment conditionality was not as serious. Parastatals can be thought of as implementing endogenous quotas that vary with domestic supply. Some countries (*e.g.* Morocco) contemplated switching to variable levies to accommodate structural adjustment conditionality until the URAA made variable levies illegal. The well-known tariff equivalence to quotas means that these different instruments can achieve nearly the same ends, as long as instrument settings can be changed as needed on a frequent basis. When a parastatal manages trade, even the revenue and welfare effects are similar, as any quota rents accrue to the government. Tariff equivalence means stocks or tax/subsidy interventions can also be used to implement price bands regimes, so stocks strategies were also used, and were more popular before structural adjustment reforms.

Several criticisms of parastatals and the use of quantitative controls rather than price measures (*e.g.* tariffs) led to elimination of these institutions in countries subject to structural adjustment reforms. Quantitative controls are less transparent that tariffs, if easier to implement. More importantly, these measures proved costly. It was believed that public marketing boards were less efficient that private marketing agents. These boards often were not only stabilising, but also subsidising specific interest groups. Their elimination may have brought greater efficiency to certain market functions (*e.g.* delivery from the farm gate to the port). Other functions that exhibited a public goods aspect (research, extension, disease protection, market information) or imperfect markets (credit and inputs) were not replaced or remained inefficient. Asian countries not subject to financial pressures have retained these institutions and the use of quantitative trade interventions (*e.g.* China and India). Cummings and Gulati (2009) argued that these were effective at stabilisation, but incurred a very high cost. These countries also held large stocks, and stocks variations were greater than import or export variations, not only during the crisis but also in many earlier years. In the face of world market extremes quantitative measures, especially complemented by stocks measures, don't face the same constraints that zero tariffs imply. Quantitative measures result in weak integration with world markets and policy interventions are focused on domestic outcomes.

How important was trade policy as a stabiliser in 2007-08?

Trade policy instruments were frequently used in a manner similar to McIntire's recommendation to stabilise and isolate domestic markets in 2007-08. But these measures in theory were simply not large enough to counteract the large world price changes being faced. Reality differed somewhat from theory, as seen in the price transmission evidence. While domestic prices increased, these increases were often substantially less than world price increases, even after tariff and tax reductions are taken into account. Moreover, stocks measures were also used frequently in combination with trade measures. Any evaluation of trade policy effects needs to take into account the extent of imperfect integration between domestic and world markets that varied across countries. While in a few cases the weak price transmission and apparent imperfect market can be explained by policy changes, in many cases explicit disconnects between world and domestic prices, beyond policy interventions, must be taken into account. Greater market integration, and more complete linkages between urban and rural prices is likely to be found where private market institutions to help stabilise farm income as well as consumption are further developed. Better market integration also allows for better sharing the localised shocks to domestic production, at both local and international levels. Weak market integration allows for greater use of stocks to manage price, however.

Trade policy measures pursued also contributed to world price instability, and to the extent to which world prices rose, at least for rice (Timmer, 2008a). The extent to which pervasive stabilisation policy across developing country importers contributed to the extent of price increases is simply not known. While it is certain that the actions of large self-sufficient traders, most rice exporters and a few large exporters contributed significantly to international instability, the combined role of small developing country importers is less well understood. A small trader closing its border may have an infinitesimal impact on world prices, but combined actions of many small traders make for more inelastic world markets. This also depends on the extent to which stabilisation achieved was due to policy versus imperfect integration. Imperfect integration also prevents a country's instability from being exported to world markets. Techniques better informed on country trade behaviour than the simple price transmission approach, used first by Tyers and Anderson (1992), are needed to assess the international effects of countries' stabilisation measures. Better models are also needed to assess just how stable world markets would be if countries kept borders open. National goals dominate in these policy decisions, so countries like China and India, which effectively stabilised in 2007-08, are unlikely to open their borders in the face of world market crisis unless a high degree of stability is expected there. There is little reason now to believe that world markets can be counted on for adequate price stability, unless the large exporters and large, self-sufficient grain producers leave their borders open. A lesson from the 2007-08 food crisis seems to be that international markets on occasion will be unreliable.

Reliable supply from world markets

High world prices mean imports to stabilise a domestic market cost more. That import levels declined very little showed that many countries are willing to bear that cost. A larger problem evident in the 1973-74 food crisis, and in the rice market in 2008, is that supplies may be unavailable from the world markets at any price. In those circumstances countries relying on international trade to stabilise or even supply domestic markets cannot achieve their domestic goals. The food security consequences of that outcome may be unacceptable. Sarris (2009) proposed an international clearing house to address this concern. But holding stocks to guarantee availability in world markets for rare events is likely to be extremely costly. What is needed is credible commitments beyond a couple of major exporters to make supplies available to importers. The United States did this in 2007-08, as farm interests objected strongly to export restrictions of the 1970s, so US legislation prevents trade embargos (IATRC, 1986). That was not sufficient to assure supplies in the world rice market, where price increases were most extreme, and the United States is only a small player.

Regional trade

Regional trade agreements may make prospects to rely on liberalised trade to stabilise domestic markets an effective option in some cases. Byerlee, Jayne and Myers (2005), using data from FAOSTAT, show that correlations of maize production across southern African countries is low (Table 4.4), so that trade could be (and is to some extent now) used to stabilise those domestic markets. But South Africa is the dominant producer, and its prices are correlated with corn prices on the Chicago Board of Trade in most years, with high transmission of those world prices to South African domestic prices (NAMC, 2009). South Africa was shifting from an import to an export position in 2007-08, so it exhibited no spike corresponding to international corn prices then (NAMC, 2009). With multiple seasons per year and irrigation, supply risks are lower for rice in Asia, but coincident weather patterns could limit effectiveness of regional trade.

In some cases, the world market for a country is its neighbour. Some of the decisions taken by rice exporters in "world markets" were because they were exporting more to neighbours. In Afghanistan, for example, imports came (at least in some regions) from Pakistan (usually a net importer) and not the large exporters to world markets. During the 2007-08 food crisis, India committed to supply Bangladesh in spite of its partial export bans, helping to offset production shortfalls in Bangladesh. Regional integration makes liberal trade policy a better stabilising option in most years. But regional trade is still at risk when world prices spike.

Open trade policy regimes in most years are likely to contribute, if imperfectly, to domestic grain market stability and food security. The policy issue is what to do during the infrequent events when world prices spike. National stabilisation goals are likely to dominate, so repeat of the policy responses in 2007-08 is very likely. Countries that can afford to do so will protect their domestic consumers. Greater reliance on trade could bring more stable world markets as well, but that approach demands that exporting

Table 4.4. **Variability and covariance of maize production in Africa, 1995-2004**

National and regional production Variability, Eastern and Southern Africa

Country/region	CV of maize production, years
Ethiopia	12.6
Kenya	8.9
Tanzania	11.2
Uganda	8.2
East Africa	**5.8**
Malawi	21.6
Mozambique	4.9
South Africa	20.3
Zambia	30.6
Zimbabwe	40.9
Southern Africa	**17.9**

Source: FAOSTAT.

Correlation matrix for maize production in Southern Africa, 1995-2004

	Zimbabwe	Zambia	Malawi	South Africa	Mozambique
Zimbabwe	1.00	0.42	0.12	0.44	–0.31
Zambia		1.00	0.09	0.38	0.11
Malawi			1.00	0.17	0.24
South Africa				1.00	0.41
Mozambique					1.00

Source: Calculated from FAOSTAT.

suppliers are reliable, even in crisis. Stocks are likely to continue to be used together with trade, especially when integration with the world market is weak. But stocks are a poor alternative to guard against very infrequent events.

4.5. Institutional arrangements

If the private sector is to play an increasing role in domestically stable grain markets, institutional development is a key element of any set of policy recommendations. There is a critical role to be played by national governments in fostering the development of institutions that are imperfect or incomplete in the least developed countries, and that need to mature as development proceeds. The role of public goods also needs to be revisited, recognising the lesson that some activities will not be taken up by private agents when the government reduces its involvement in grain markets. The difficult problem of how to phase reforms and liberalisation is resolved only slowly as private sector institutions develop. Some of the World Bank's (Byerlee, Jayne and Myers, 2005) "best practices" recommendations are best seen as a longer term goal unlikely to be achieved quickly following reforms. The two issues briefly examined here are what institutions need to develop in light of stabilisation goals of governments, and what is the appropriate role of government as that development proceeds?

Market institutions

Private sector institutions are needed for long run agricultural development, better marketing and distribution, input provision, storage, credit and risk management. Public sector institutions must provide public goods, possibly in different ways than in the past, including market information, research, extension, and infrastructure. Private institutional development is also likely to require public sector actions to foster institutional development. For example, new legal frameworks may be needed to permit credit and insurance schemes to develop.

Parastatals and reforms

Following structural adjustment reforms and privatisation of parastatal grain marketing boards, at one point recommendations were that the government should withdraw almost completely from grain markets. Few activities were recognised as public goods that a government must provide. A prevailing view was that in many instances government did more harm than good. Even provision of market information was viewed for a time as something to be addressed as a largely a private sector activity, at least by some foreign aid donors. Subsequently, it has become clear that the private sector quickly takes on certain activities, but some activities have a sufficient public goods nature that they are not undertaken by private agents. The private sector readily takes over basic marketing and distribution functions, such as sales to or purchases from international markets, but the efficiency of those activities depends on the state of infrastructure and legal institutions. Activities likely to need government assistance or direct provision that are related to storage and stabilisation include both basic infrastructure, such as roads, and legal and institutional "infrastructure" such as market information and legal frameworks that facilitate private marketing and storage. Warehouse receipts are an often cited example that enables commercial storage and trade by allowing transactions to take place without the commodity physically moving. Market information, another critical element, includes not only insuring widespread and accurate price information at a point in time, for more efficient distribution, but also early warning on production or international market events. Timely information throughout the crop year helps not only commercial trade but also decisions by farmers. Legal institutions were unnecessary when government physically handled domestic trade, and must advance if commercial storage mechanisms are to develop. Credit and finance are also necessary for both storage and input markets to develop, and have proven slow to evolve after privatisation of agricultural markets.

Some of these examples, like market information, are clearly public goods that government must provide, while others will become private if marketing institutions develop. Stabilisation is not the only objective that necessitates these reforms, but they are necessary not sufficient conditions for better integrated markets, especially within the domestic market. These changes are not about stabilisation policy per se, but bringing commercialisation and more integrated markets, allowing private storage to play a more active role and reducing intervention needed by governments. They may also make interventions more widely effective.

In the wake of structural adjustment reforms the World Bank (CRMG, 2008) has promoted development of private arrangements to enable farmers and traders to manage risk, including forward pricing and crop insurance. These were initially seen as a substitute for public stabilisation. Countries and developing country farmers have been quite slow to adopt these institutions, however (CRMG, 2008; Galtier, 2009a and 2009b). In many cases domestic marketing institutions have been immature, preventing adoption of these approaches.

Futures markets

Scale and transaction costs were recognised early on as issues in the use of futures markets as a stabilisation tool for developing country farmers. Since the size of contracts on the Chicago Board of Trade (CBOT) is large relative to production by small scale farmers, and especially poor farmers of less than two hectares, intermediaries were needed to create contracts of appropriately small size for farmers. It was feared that transaction costs as a fraction of price could be quite large for scaled down contracts, and few examples of successful intermediaries arose. Co-operatives were one approach proposed, but these largely political institutions (in Africa) have been ineffective trading agents in many cases, and were not ready to take on a role as contractor for insuring or stabilising transactions. The weak legal framework and imperfect financial markets post structural adjustment also made creation of institutions to scale down contracts for farmers problematic. Most discussions with the World Bank on implementing this approach ended up being with public institutions (CRMG, 2008).

Even in developed countries, farmers have been reluctant to utilise futures markets (Carter, 1999). The time frame of contracts for which futures markets are not thin is of a short duration. It is more reliable to hedge for a few months than across years. The number of contracts held for a long duration is much smaller than for an upcoming harvest, and there is relatively little information about future prices in those long-term contracts, beyond information on price for the upcoming or recent harvest and storage costs from that. Farmers cannot effectively stabilise prices across years, even in developed countries, and they make limited use of forward sales at planting. Futures markets are indispensible tools for the large traders, especially on international markets (*e.g.* Cargill and ADM), who want to minimise price risk as they buy, hold or transport, and then sell at a later time but collect only small margins. But forward pricing for farmers is often through intermediaries who are traders and who have storage (*e.g.* grain elevators), and who engage in futures market transactions on farmers' behalf. The World Bank recognised after a decade of effort on this approach that futures and options are better suited to traders than farmers (CRMG, 2008). In the longer run, as countries develop and markets become more commercial, traders will offer forward pricing options to farmers. But developed country experience suggests even then these will be for transactions within a marketing year and not as an inter-annual stabilising tool.

Byerlee, Jayne and Myers (2005) argued that futures and options transactions should be practiced by private agents, not government. They noted that trader capacity to utilise these institutions is very low in Africa, however. Traders in only a few countries have utilised this option, but new futures markets have actually arisen in China, India and South Africa. Basis risk means prices on the CBOT may not be highly correlated with domestic prices. If traders or farmers hedge on the CBOT, that may bear no relation to the revenues or costs they realise on their farming or trading operations. For local traders, forward pricing is needed on the prices they actually face, which may differ from world prices. CBOT prices may be more appropriate for import costs or export revenue than for domestic price stabilisation, especially if domestic volatility drives domestic prices in most years. Moreover, where local futures markets have developed recently (in China, India and South Africa), domestic markets are very large and domestic trading institutions are more mature than in most developing countries. It appears that futures markets work better in these settings when the markets are those of the country in question, and large scale needs to be achieved for these new institutions to arise.

As traders evolve, so does commercial storage. As marketing institutions mature, forward pricing means more efficient domestic storage. Thus, policies to assist in development of institutions to take advantage of forward pricing, and to facilitate commercial trading, are advantageous, but only become options as marketing systems mature.

Sarris (2009) has recently proposed the use of futures and options as a means by which developing countries can minimise foreign exchange risk. He has shown that countries would have benefited from use of forward pricing during the recent food crisis, keeping foreign exchange costs under control and more predictable in the face of rising world food prices. But the key issue is who will engage in those transactions? Is a parastatal entity necessary or can some other institutional arrangements exist? Sarris does not elaborate a private institutional framework, so the government remains the logical institution to benefit from this approach. The World Bank has pursued this approach with governments, but with only a couple of decisions taken to actually buy futures contracts (CRMG, 2008), and there have been some rejections of proposals.

Crop insurance

There is more interest now in crop insurance for developing countries, the risk management/stabilisation alternative the World Bank now emphasises (CRMG, 2008). A major issue in design and implementation of crop insurance is moral hazard. That is, there needs to be a payoff criterion that is relevant to a farmer's operation, but which does not permit shirking or failure to pay for inputs so that lower yields result in a pay-out. It is difficult for insurers to get reliable information to use to gauge when pay-outs on policies should occur, as well. Pay-out criteria are set based on weather indices or outcomes over large areas, where outcomes may not coincide with on-farm outcomes. Moreover, only farmers at high risk may choose to purchase insurance, the problem of adverse selection. Basis risk, that local prices may not be adequately correlated with index prices, and that in immature markets price information may be poor, complicate the design of crop insurance for developing country farmers, as well. Design of insurance contracts that avoid moral hazard and adverse selection issues have also been problematic for insurance approaches in developed countries (Anton, 2009). Farmers in the United States complain, for example, that pay-outs based on state-wide yields do not result in pay-outs when their own output is low. Scale and transaction costs to serve small farmers are also issues for crop insurance.

That bad weather is not an isolated local event, harming harvests nationally, may mean losses of many farmers occur in the same year and so insurers face large simultaneous pay-outs. Crop insurance alternatives better protect against random domestic sources of volatility than against systematic events, like world price changes that affect all farmers, as well as national droughts. In systematic events, pooling of risk across farmers is not an option. Reinsurance of those local contracts on international insurance markets is essential and may be costly. In the event of an international food crisis, demands on the reinsurance pool would coincide as well.

Experience with crop insurance in developed countries should not be seen as encouraging. Like parastatal price stabilisation, crop insurance programmes have often incorporated implicit subsidies. They have either raised mean prices or benefited some types of farmers at the expense of others. When there are no subsidies, demand for insurance by farmers has been low. Crop insurance schemes have also been part of government programmes and have not existed to any significant extent as an entirely private enterprise (Anton, 2009).

Both forward pricing and crop insurance are risk management tools that develop as agriculture develops and becomes commercialised. Benefits will arise to farmers, and markets will become more stable as these institutions evolve. The primary beneficiaries will be traders and commercial storage operators, with benefits to farmers following. Until markets mature and commercialisation advances, it is unlikely that local traders will utilise existing futures markets to help stabilise farm gate prices or incomes. In most developing countries forward pricing options are still a premature alternative to substitute for the stabilisation policies that existed before reforms. Crop insurance may be a better option in the shorter term, but designing affordable contracts remains quite difficult, and when markets work well insurance protects farmers against domestic events, not world price spikes. While development of these institutions should be encouraged, they are not tools that substitute for market stabilisation policies in poor countries nor do they protect against infrequent world price spikes as experienced during the 2007-08 food crisis.

Governance

The key to effective private sector participation in any stabilisation strategy, as in any successful agricultural development strategy, is effective governance. A lesson from the aftermath of the privatisation era is that some role for government remains, including provision of public goods and fostering institutional development, including fostering development of new private market institutions.

Parastatal reform and coexistence

Following structural adjustment reforms, parastatal marketing boards were eliminated in many developing countries. Reform conditionality had a much bigger effect on policy than did trade liberalisation initiatives, such as the URAA agreement (Abbott, Andersen and Tarp, 2010). It was driven in part by the financial losses of those entities. Problems arose out of stabilisation efforts, but were exacerbated by policies that subsidised as well as stabilised. Stabilisation objectives persisted after reforms, and contributed to lack of country "ownership" of privatisation reforms. Countries under less pressure from the IMF and World Bank, principally in Asia, did not undertake these reforms, and one still finds parastatal grain boards functioning there. Those boards effectively stabilised during the 2007-08 food crisis, at a cost. Somewhat surprisingly, parastatals have also persisted in eastern Africa (Jayne and Jones, 1997).

While parastatal grain marketing institutions may continue to exist outside Asia, external pressures have led to changes in the way they operate in many cases. In Morocco, for example, the marketing board ONICL no longer physically handles international grain trade, but still manages international tenders for its wheat imports. While private traders conduct trade, ONICL retains some control on the magnitude and timing of imports. Trade policy clearly and consistently delineates stabilisation goals there. In eastern Africa, and elsewhere, it is not uncommon for parastatals and private traders to coexist in various ways. This compromise has been pushed in both structural adjustment and WTO reform agendas, even in China. Jayne and Tschirley (2009) illustrate some problems that can arise. If incentives and signals to private traders are inconsistent or opaque, they may not act in a timely manner to achieve food security. Rules can be stacked against the private sector, so they may not take up functions expected after the reforms. Response to a crisis is likely to elicit both the need for government intervention, and conflicts between public and private trade interests, as Jayne and Tschirley's examples illustrate. For coexistence of public and private trade to work, there must be clear delineation of regulations and assignment of functions. If trade liberalisation is the norm, to dictate institutions and behaviour in normal years, but governments must intervene in crises, then clear and transparent rules must characterise that intervention and when it will occur.

Developing new private market institutions

Countries that eliminated parastatal marketing boards often also saw problems in the provision of credit, affecting marketing and distribution as well as input provision. In addition, private sector trade requires different legal frameworks that were seldom implemented upon reform. If the private sector is to play an increasing role, but infrequent intervention by governments will continue to avoid catastrophes, institutions that facilitate private storage and marketing need to be reformed, as well. These are critical to both long run agricultural development and short run adjustments to changing market conditions.

Governance problems can arise from inconsistencies and corruption. They may simply reflect national objective functions that put low weight on agriculture, however. The disconnect between international donor and developing country government responses to the 2007-08 food crisis is mostly due to differing objectives, hence different priorities. This can spill over to international solutions. Gilbert (1996) argued that the fundamental reason behind the failure of international commodity agreements was the failure of participating governments to agree on common objectives. The risk with regional trade solutions also lies in potentially differing objectives of trade agreement partners. If a country within a free-trade area wants to stabilise, its actions will spill over onto its trade partners, who may not want the same degree of stabilisation or at the same price level.

The cases described by Jayne and Tshirley (2009) and by Poulton *et al.* (2006) illustrate the importance of overcoming political failures if stabilisation and broader food security objectives are to be achieved. In the cases where grain market stabilisation succeeded in the face of food crisis, credible and transparent policy prevailed. The problem cited by economists with these cases is always high cost. Government commitment, another key to success, means that government was willing to pay those costs. It is not necessary, however, to return to public institutions managing grain markets if the governance problems with coexistence and commitment are solved.

4.6. Conclusions

During the 2007-08 food crisis dramatic world grain price increases brought stabilising policy responses by many developing country governments. The isolationist policies pursued by governments contradicted existing "best practices" risk management strategies that focus on long run agricultural development, trade liberalisation, safety nets and private market solutions to risk (Byerlee, Jayne and Myers, 2005). Domestic market outcomes were conditioned to varying degrees by lagged, imperfect price transmission, transaction costs and weak market integration in addition to policy. Stabilisation of domestic markets also spilled over into greater international market instability. Countries that had opened their borders were vulnerable to high import costs and pass-through to high consumer prices, which was estimated to have brought hunger, malnutrition and poverty to an additional 100 million people (FAO, 2008b; Rosen *et al.*, 2008; World Bank, 2008a).

Trade policy responses to future world price spikes are likely to look much like the responses to the just past food crisis. National priorities reflect persistent stabilisation objectives and broad concern with consumer welfare. Governments alter tariffs or quantitative controls as the infrequent price spikes are realised, acting like a price bands regime. While these responses spill over into greater international market instability, unless the very large exporters and large, now self-sufficient, markets (*e.g.* China and India) leave their borders open, it is unlikely that liberal trade will result in sufficient stability in world prices to allow most developing country importers to leave their borders open. Combined actions by small importers may also make world markets less elastic, even if individual country policies have a very small influence on world prices. Asian parastatals maintained quite stable domestic markets during the 2007-08 food crisis using this regime and experts doubt they would behave otherwise in the future.

Galtier (2009a and b) and others have argued that policy recommendations on risk management, and the de-emphasis on domestic price stabilising regimes, are no longer appropriate in light of the food crisis. Moreover, recommended private risk management institutions, including forward pricing and crop insurance, had seldom materialised following structural adjustment reforms. A new perspective on policy must re-think "best practices" for both risk management and stabilisation in light of the infrequent but real world price spikes that require combating international and domestic sources of volatility in those years.

Policy recommendations

Policy recommendations follow from objective functions of policy makers, which need to be clear on the priorities given to agriculture, food security, stability and poverty. In spite of estimates by economists of small benefits to stability (Byerlee, Jayne and Myers, 2005), developing country governments have shown a preference for stability both in their reluctance to adopt reforms following structural adjustment and in their responses to the food crisis. The international donor community had emphasised the use of safety nets to protect the extreme poor and promotion of long-run agricultural development, reflecting an objective function putting heavier political weight on prevention of extreme rural poverty and on longer run outcomes. The short-run focus of developing countries reflects concern for short run stability and broader consumer protection. National policy responses may also reflect a desire to avoid catastrophes and extremes, by pursuing policy regimes that try to avoid consequences of being in the tails of price distributions.

Policy responses should be conditioned by future expectations on world price distributions versus domestic sources of variability. Existing recommendations were based on a presumption of stable world markets and the dominance of production variability as a contributor to domestic instability. Policy makers and analysts now need to recognise that there will be episodes of high, volatile world prices and of low, stable world prices, driven by external factors. The level, variability and even predictability of agricultural commodity prices are now conditional on circumstances both internal and external to those markets. Policy makers must pay attention to related markets that have caused world price increases in the past, aware that new mechanisms arise. They should expect infrequent but large spikes when relying on international markets to smooth domestic markets. It is likely that we may return to a period of stability like 1998 to 2005, where domestic factors dominate, but that spikes in world grain prices can reoccur. Macroeconomic, financial market and energy market factors will influence this future world price distribution.

Since domestic sources of instability do dominate in most years, policy must first address that domestic volatility. Liberal trade policy works well toward that end in those years when world prices are low and stable. The "best practices" recommendation combines trade liberalisation with self-sufficiency, to use trade as a stabilising mechanism, but not letting the share of imports become excessive due to the neglect of domestic agriculture. Early warning of both domestic shortfalls and world price trends is needed if there is to be greater reliance on trade, however, necessitating improved market information systems. The biggest risks from relying on trade, however, are that world markets fail during crises, as the rice market did in 2008, and supplies are unavailable at any price.

Public stocks management as an alternative to trade policy has been seen as a costly option, since large stocks must be carried for long periods if crises are infrequent. The use of trade alternatives requires that financial resources allow imports when needed. It also requires that world markets are reliable suppliers.

Imperfect integration into world markets and seasonal price dynamics along with delays in import delivery mean that short-run, inter-seasonal stockholding policy will complement trade policy to implement a stabilising regime. Contrary to theory that looks only at annual carry-overs, stocks management is used along with trade adjustments in most developing countries to adjust to both production shortfalls and world price spikes. Seasonality and short run price dynamics also mean stocks will always exist, and carry-out stocks will smooth both current and future consumption. Trade and stocks adjustments must also prepare a country for repeated shortfalls, as imports replace depleted stocks. Stocks management can also prevent domestic price spikes in critical pre-harvest periods, before imports arrive.

Improved private domestic risk management institutions will facilitate better stocks management as well as greater reliance on trade. In the least developed countries imperfect legal frameworks, market information systems, and financial institutions impede development of commercial stocks as well as the ability of trade or stocks strategies to broadly protect domestic consumers. Stocks need not be publicly held if consistent and transparent rules govern stocks management. Better private institutions mean smaller stocks and more effective interventions.

The aftermath of structural adjustment reforms has demonstrated that a role for government remains in assuring food security and developing agricultural markets. While private markets may more efficiently provide most marketing services, public goods require government provision and the government must foster institutional development, public and private. While the cost of maintaining parastatal marketing boards is likely to be high, they were effective in stabilising and avoiding increased poverty in many Asian markets. Elsewhere stabilisation policy is pursued by a new mix of public and private trade. Consistency, transparency and predictability of interventions is key if the private sector is to play a role in stabilisation, and especially if public and private trade are to coexist. Governance failures can exacerbate food security crises, whether of domestic or international origin.

While the agricultural trade policy regime should rely on liberal trade in most years, it should be recognised that short run dynamics mean short-run stocks policy remains a viable concern, due to delays in import arrival and inter-seasonal price dynamics. Moreover, trade policy adjustments are likely to be necessary when infrequent world price spikes reoccur. The challenge to implementing such a regime is that consistent, predictable and transparent governance is needed so that interventions make outcomes better, not worse.

Future research agenda

World grain price spikes have been short run phenomena eliciting short run policy responses. Many of the anecdotes about market failures and governance failures were more about poorly timed reactions or short-term information inadequacy than about incorrect eventual levels of intervention. The extent of problems becomes masked in annual average data, as well. The dynamics of storage and of price transmission require that trade and stocks policy be considered under that shorter term time frame. Most of the work on stabilisation policy has utilised annual models to examine stocks and trade options, however. Data requirements and availability mean that seasonal price dynamics would be difficult to model, but stylised assumptions based on cost of storage theory and presumptions of inelastic demand would enable examination of these issues under shorter time frames. One element of that work should be to understand better how private trading evolves in developing countries, and how marketing institutions develop to change the roles of private commercial storage versus public storage. Another would be to better understand why the lags in price transmission occur as they do, and how that is affected by policy in place. The benefits of developing better market information, including better early warning systems, could be compared to the alternatives of building better port and distribution infrastructure to deliver imported food more quickly to needy populations. Benefits to forward pricing options could also be better estimated under the appropriate (short-run) time frame for those options. Rules on stockholding and on trade flows could be explored in that framework to compare how costly these alternatives actually are in light of dynamic market adjustments. A short run model would also provide better insight into the extent to which world price spikes actually are passed to consumers, and the effects that might have on short run nutritional status and poverty. It is likely that better insight into stabilisation policy would emerge if research looks more closely at the time frame under which it operates.

A related issue critical to the analysis of this chapter is our understanding of the distribution of world grain prices. The assertion here is that this price distribution, and its uncertainty, is conditional on external factors and exhibits infrequent but large spikes. Policy recommendations are conditional on the shape of those world price distributions. Much of the work subsequent to the 2007-08 food crisis has utilised time series methods that assume no structural changes in this distribution, and do not capture any irregularities in the shape of that distribution, however. Better understanding of the historical distribution of world grain prices and the factors driving those prices will better inform assumptions on future price distributions and so implications of those distributions for policy design. This will require that simple assumptions on symmetric normal distributions be abandoned, and that policies be considered in light of the likely world price distribution that may occur in the future. Risk management and stabilisation policy strategies should be evaluated under more realistic assumptions on these world price distributions.

The effectiveness of stabilisation policy is likely dependent on the state of marketing institutional development. One of the "best practices" recommendations that stands is that improvement in these private market institutions will help farmers, consumers and governments manage risk. The state of these institutions is immature in most developing countries. Moreover, while it was eventually recognised that market liberalisation reforms should be phased in, how to implement that slowly, and even what that meant for policy evolution, were unknown. In addition, recommendations often resulted in coexistence of both public and private trade, but with unclear divisions of labour between the public and private sectors. Research should develop a better understanding of how marketing institutions can and should develop, and what their development may mean for development of risk management strategies. It should lead to a better understanding of the conditions under which public marketing boards and private trade can coexist, and what regulations lead to better functioning markets. It should inform policy makers on what market conditions and what stage of development is needed for some of the more advanced marketing institutions to work. More realistic assumptions on when more sophisticated risk management institutions might work are needed by both policy makers and researchers.

The focus of this chapter has been on domestic stabilisation options, but domestic policy choices both depend on and create problems for international stabilisation strategies. A well-known issue is that domestic stabilisation spills over into greater international market instability, and the extensive stabilisation by countries in 2007-08 contributed to the height of the price peaks realised. The extent to which world markets would have been stable and how high prices might have risen had countries left their borders open, however, is uncertain. Methods to model this have utilised very simple assumptions on price transmission or imperfect domestic-foreign good substitutability, typically assuming any blockage of international price transmission is the consequence of policy not weak market integration. The extent of world market stability should be explored using better estimates of price transmission behaviour, and better models than simple price transmission and the law of one price to capture border interventions (such as average cost pricing by local traders or models that assume parastatals react to quantity and not price signals). This is also another case where shorter term models might better inform the extent of world price instability, taking into account the observed lags in price transmission. Price transmission and the stability of world prices, as well as options to use trade to stabilise domestic markets, depend on whether any of the proposed international mechanisms, such as a virtual reserve or clearinghouse, are adopted. Such research should look at the worldwide consequences of alternative assumptions on domestic stabilising regimes by a country, its trade partners and other important world market actors, including regimes that only protect against being in the tails of price distributions.

In general, the interaction between domestic and international markets needs to be better understood and better modelled so that evaluation of domestic policy options is better informed by realistic assumptions on world market behaviour and on the international stabilisation mechanisms adopted or proposed, as well as better assumptions on how domestic markets behave.

Note

1. The author is grateful for comments received from Jonathan Brooks and Jesús Antón, and from participants at the 2010 OECD Global Forum. The views expressed and any errors or omissions are solely those of the author and should not be attributed to the OECD.

References

Abbott, P. (2009), "Development dimensions of high food prices", *OECD Food, Agriculture and Fisheries Working Paper*, No. 18, OECD, Paris.

Abbott, P., T.B. Andersen, and F. Tarp (2010), "IMF and Economic Reform in Developing Countries", *The Quarterly Review of Economics and Finance,* Vol. 50, 17-26.

Abbott, P. and A. Borot de Battisti (2011), "Recent global food price shocks: causes and consequences for the developing world", *Journal of African Economies,* Vol. 20, pp. i12-i62.

Abbott, P., P. Konandreas and M. Benirschka (1993), "A Model for Assessing Food Security Policy Alternatives", in P.B.A.D. Bigman (ed.), *Food security and food inventories in developing countries,* CAB International, Wallingford, UK.

Abbott, P.C., C. Hurt and W.E. Tyner (2008), *What's driving food prices?* Farm Foundation Oak Brook, United States.

Abbott, P.C., C. Hurt and W.E. Tyner (2009), *What's driving food prices?* March 2009 update, Farm Foundation? Oak Brook, United States.

Abbott, P.C., P. Patterson and L. Young. (1998), "Plans and adjustment: A structural approach to modeling grain importer behavior", in T. Yildirim, A. Schmitz and W.H. Furtan (eds.), *World agricultural trade: Implications for Turkey*, Westview Press, Boulder, CO.

Abbott, P.C., W.E. Tyner and K. Cripe (1996), The GATT outcome and price stability: World wheat markets and Moroccan import policy, *Interconference Symposium on "GATT Implementation and Structural Adjustment in the Mediterranean Region"*, Rabat, Morocco, Moroccan Association of Agricultural Economy and the International Association of Agricultural Economists.

Anderson, K. and W. Masters (eds.) (2009), *Distortions to agricultural incentives in Africa*, World Bank Washington, DC.

Anton, J. (2009), "Risk management in agriculture: A holistic conceptual framework", in OECD, *Managing risk in agriculture: A holistic approach*, Paris, OECD.

Arndt, C., R.B. Nelson, M. Antonio, M.D. Nucifora and J.T. Thurlow (2008), "Higher fuel and food prices: Impacts and responses for Mozambique", *Agricultural Economics*, Vol. 39, pp. 497-511.

Aube, T. (2005), "Improving the methodology for Joint FAO/WFP crop and food supply assessment missions: Estimating commercial imports", *Technical Consultation on Crop and Food Supply Assessment Missions (CFSAMS)*, Rome, FAO.

Badaine, O. and D. Resnick (2005), "Regional trade liberalization and domestic food market stabilization in African countries", *Managing Food Price Instability in Low-Income Countries*, World Bank Washington, DC.

Baffes, J. and B. Gardner (2003), "The transmission of world commodity prices to domestic markets under policy reforms in developing countries", *Policy Reform*, Vol. 6, pp. 159-180.

Balcombe, K. (2009), "The nature and determinants of volatility in agricultural prices: An empirical study from 1962-2008", *Workshop on Institutions and Policies to Manage Global Market Risks and Price Spikes in Basic Food Commodities*, FAO, Rome.

Balcombe, K., A. Bailey and J. Brooks (2007), "Threshold effects in price transmission: The case of Brazilian wheat, maize, and soya prices", *American Journal of Agricultural Economics*, Vol. 9, pp. 308-323.

Bale, M. D. and E. Lutz (1979), "The effects of trade intervention on international price instability", *American Journal of Agricultural Economics*, Vol. 61, pp. 512-516.

Barrett, C.B. (2001), Measuring Integration and Efficiency in International Agricultural Markets, *Review of Agricultural Economics*, Vol. 23, pp. 19-32.

Barrett, C.B. and D.G. Maxwell (2005), *Food Aid After Fifty Years: Recasting its Role*, Routledge, London.

Berck, P. and D. Bigman (1993), *Food Security and Food Inventories in Developing Countries*, CABI International, Wallingford, United Kingdom.

Blandford, D. (1983), "Instability in World Grain Markets", *Journal of Agricultural Economics* Vol. 43, pp. 379-395.

Boussard, JM., F. Gérard, M.G. Piketty, A.K. Christensen and T. Voituriez (2004), "May the pro-poor impacts of trade liberalisation vanish because of imperfect information?" *Agricultural Economics*, Vol. 31, pp. 297-305.

Boussard, J.-M., F. Gérard, M.G. Piketty, M. Ayouz, and T. Voituriez (2006), "Endogenous risk and long run effects of liberalization in a global analysis framework", *Economic Modelling*, Vol. 23, pp. 457-475.

Braverman, A., R. Kanbur, A.S.P. Brandao, J. Hammer, M.D.R. Lopes and A. Tan (1992), *Commodity price stabilization and policy reform*, World Bank, Washington, DC.

Brooks, J. (2009), *Agricultural Policy Design in Developing Countries: Interim Report*, OECD, Paris.

Brooks, J. and J. Dewbre (2006), "Global trade reforms and income distribution in developing countries", *Journal of Agricultural and Development Economics*, Vol. 3, pp. 86-111.

Brooks, J., G. Dyer and E. Taylor (2008), "Modelling agricultural trade and policy impacts in less developed countries", *OECD Food, Agriculture and Fisheries Working Papers*, No. 11, OECD, Paris.

Brooks, J. and O. Melyukhina (2005), "Estimating the pass-through of agricultural policy reforms: An application to Brazilian commodity markets", *OECD Food, Agriculture and Fisheries Working Papers*, No. 2, OECD, Paris.

Buccola, S.T. and C. Sukume (1988), "Optimal grain pricing and storage policy in controlled agricultural economies: Application to Zimbabwe", *World Development*, Vol. 16, pp. 361-371.

Byerlee, D., T.S. Jayne and R. Myers (2005), *Managing food price risks and instability in an environment of market liberalization*, World Bank, Washington, DC.

Cafiero, C., E.S.A. Bobenrieth, J.R.A. Bobenrieth and B.D. Wright (2011), "The empirical relevance of the competitive storage model", *Journal of Econometrics*, Vol 162, pp. 44-54.

Carter, C. (1999), "Commodity futures markets: A survey", *The Australian Journal of Agricultural and Resource Economics*, Vol. 43, pp. 209-247.

Commodity Risk Management Group (CRMG) (2008), *The International Task Force on Commodity Risk Management in Developing Countries: Activities, findings and the way forward*, World Bank, Washington, DC.

Cudjoe, G., C. Breisinger and X. Diao (2008), Food price transmission and poverty impacts in Ghana, *Discussion Paper* No. 842, IFPRI, Washington, DC.

Cummings, R. and A. Gulati (2009), "Grain price stabilization experiences in Asia: What have we learned?" in *Workshop on institutions and policies to manage global market risks and price spikes in basic food commodities*, FAO, Rome.

Dand, R. (1999), *International cocoa trade*, Second edition, Woodhead Publishing, New York.

Daviron, B., M. Aubert, N. Bricas, H. David-Benz, S. Dury, J. Egg, F. Lançon and V. Meuriot (2008), *Les mécanismes de transmission de la hausse des prix internationaux des produits agricoles dans les pays africains*, Fondation pour l'Agriculture et la ruralité dans le monde (FARM), CIRAD, Paris.

Dawe, D. (2008), "Have recent increases in international cereal prices been transmitted to domestic economies? The experience in seven large Asian countries", *ESA Working Paper* No. 08-03, FAO, Rome.

Dawe, D. and C. Morales-Opazo (2009), *How much did developing country domestic staple food prices increase during the world food crisis? How much have they declined*, FAO, Rome.

Dawe, D. and C.P. Timmer (2005), "Managing food price instability in Asia: A macro food security perspective", in *Managing Food Price Instability in Low-Income Countries*, World Bank, Washington, DC.

Deaton, A.S. (1992), "Commodity Prices, Stabilization, and Growth in Africa", *Research Program for Development Studies Discussion Paper 166*, Princeton University, Princeton, NJ.

Delgado, C. (2009), "Driven by increased price volatility, the global food crisis is not over", in OECD (2009), *Policy Dialogue on High Food Prices: Outlook and Donor Mid-term Responses*, OECD, Paris.

Delgado, C. and R. Townsend. (2009)," Managing food price volatility: National options in small, poor countries", *Institutions and Policies to Manage Global Market Risk and Price Spikes in Basic Food Commodities*, FAO Experts Consultation, Rome.

Demeke, M., G. Pangrazioand M. Maetz (2008), *Country responses to the food security crisis: nature and preliminary implications of the policies pursued*, FAO, Rome.

Denning, G.E.A. (2009), Input subsidies to improve smallholder maize productivity in Malawi: Toward an African green revolution, *PLoS Biology*, Vol. 7, pp. 2-10.

Dewbre, J., C. Giner, W. Thompson and M. Von Lampe (2008), "High food commodity prices: Will they stay? Who will pay?" *Agricultural Economics*, Vol 39, pp. 393-403.

Diouf, J. (2008), *High food prices: Causes and possible actions*, FAO, Rome.

Dorin, B. (2003), *From Ivorian cocoa bean to French dark chocolate tablet*, CIRAD, Paris.

FAO (2009), *GIEWS national basic food prices*, FAO, Rome.

FAO (2008a), "Global market analysis: Special feature on volatility in agricultural commodities", *Food Outlook*, FAO, Rome.

FAO (2008b), *The State of Food Insecurity in the World: 2008*, FAO, Rome.

Galtier, F. (2009a), "How to manage food price instability in developing countries?" CIRAD, UMR MOISA, Montpellier, France.

Galtier, F. (2009b), "The need for public intervention to stabilise prices", *Perspective: Food Security*, CIRAD, Montpellier, France.

Gilbert, C.L. (1996), "International commodity agreements: An obituary notice", *World Development*, Vol. 24, pp. 1-19.

Gilbert, C.L. (1997), *Cocoa market liberalization: Its effects on quality, futures trading and prices*, The Cocoa Association of London, London.

Gilbert, C.L. (2009), "How to understand high food prices", *Institutions and policies to manage global market risk and price spikes in basic food commodities*, FAO, Rome.

Haq, Z.U., H. Nazli and K. Meilke (2008), "Implications of high food prices for poverty in Pakistan", *Agricultural Economics*, Vol. 39, pp. 477-484.

Hayami, Y. and K. Anderson (1986), *The political economy of agricultural protection*, Allen and Unwin, London.

Hazell, P., G. Shields and D. Shields (2005), "The nature and extent of domestic sources of food price instability and risk", *Managing food price instability in low-income countries*, World Bank, Washington, DC.

Headey, D. and S. Fan (2008), Anatomy of a crisis: the causes and consequences of surging food prices, *Agricultural Economics*, Vol. 39, pp. 375-391.

Huddleston, B., D.G. Johnson, S. Reutlinger and A. Valdes (1984), *International finance for food security*, Johns Hopkins University Press, Baltimore.

IATRC (1986), "Embargoes, surplus disposal and US Agriculture", *Agricultural Economic Report*, Economic Research Service, USDA Washington, DC.

International Monetary Fund (IMF) (2010), *International financial statistics*, IMF, Washington, DC.

Irwin, S., D. Sanders and R. Merrin (2009), "Devil or angel: The role of speculation in the recent commodity price boom (and bust)", *Southern Agricultural Economics Association Meetings*, Atlanta, Georgia.

Irwin, S.H., P. Garcia, D.L. Good and E.L. Kunda (2008), "Recent convergence performance of cbot corn, soybean, and wheat futures contracts", *Choices*, Vol. 23, pp. 16-21.

Ivanic, M. and W. Martin (2008), "Implications of higher global food prices for poverty in low-income countries", *Agricultural Economics*, Vol. 39, pp. 405-416.

Jayne, T.S., J. Govereh, A. Mwanaumo, J.K. Nyoro and A. Chapoto (2002), "False promise or false premise? The experience of food and input market reform in eastern and southern Africa", *World Development*, Vol. 30, pp. 1967-1985.

Jayne, T.S. and S. Jones (1997), "Food marketing and pricing policy in eastern and southern Africa: A survey", *World Development*, Vol. 25, pp. 1505-1527.

Jayne, T.S. and D. Tschirley (2009), "Food price spikes and strategic interactions between the public and private sectors: Market failures or governance failures?", *Institutions and Policies to Manage Global Market Risks and Price Spikes in Basic Food Commodities*, FAO, Rome.

Jones, D. and A. Kwiecinski (2010), "Policy responses in emerging economies to international agricultural commodity price surges", *OECD Food, Agriculture and Fisheries Working Papers*, No. 34, OECD, Paris, doi: *10.1787/5km6c61fv40w-en.*

Just, D.R. and R.E. Just (2008), "Monopoly power, futures market manipulation, and the oil price bubble", *Journal of Agricultural and Food Industrial Organization*, Vol. 6, Article 2.

Kannapiran, C. (2000), "Commodity price stabilization: macroeconomic impacts and policy options", *Agricultural Economics*, Vol. 23, pp. 17-30.

Kherallah, M.C. Delgado, E. Gabre-Madhin, N. Minot and M. Johnson (2002), *Reforming agricultural markets in Africa*, Johns Hopkins University Press, Baltimore, MD.

Lence, S.H. (2009), Do futures benefit farmers? *American Journal of Agricultural Economics*, Vol. 91, pp. 154-167.

Loening, J.L., D. Durevall and Y.A. Birru (2009), *Inflation dynamics and food prices in an agricultural economy: The case of Ethiopia*, unpublished.

Lustig, N. (2009), *Coping with rising food prices: Policy dilemmas in the developing world*, Center for Global Development, Washington, DC.

Macbean, A.I. (1966), *Export instability and economic development*, Harvard University Press, Cambridge, MA.

Mcintire, J. (1981), "Food security in the Sahel: Variable Import Levy, Grain Reserves, and Foreign Exchange Assistance", *Research Report 26*, International Food Policy Research Institute (IFPRI), Washington, DC.

Mercoiret, M.R. and J.M. Mfou'Ou (2006), *Rural Producer Organizations (RPOs), empowerment of farmers and results of collective action*, Rural Producers Organizations for Pro-poor Sustainable Agricultural Development, Paris, World Bank.

National Agricultural Marketing Council (NAMC) (2009), *The South African food cost review: 2008*, Pretoria, National Agricultural Marketing Council, Republic of South Africa.

Negassa, A. and T.S. Jayne (1997), *The response of ethiopian grain markets to liberalization*, Ministry of Economic Development and Co-operation, Addis Ababa, Ethiopia.

Newberry, D.M.G. and J.E. Stiglitz (1981), *The theory of commodity price stabilization*, Oxford University Press, London.

OECD and FAO (2008), *OECD-FAO Agricultural Outlook 2008-2017*, OECD, Paris.

Pardey, P. G. and J.M. Alston (2010), "US agricultural research in a global food security setting", *A Report of the CSIS Task Force on Food Security*, Washington, DC, Center for Strategic and International Studies.

Pinckney, T.C. (1993), "Is market liberalization compatible with food security? storage, trade and price policies for maize in Southern Africa", *Food Policy*, Vol. 18, pp. 325-333.

Platteau, J.P. (1991), "Traditional systems of social security and hunger insurance: past achievements and modern challenges", in E. Ahmad, J. Dreze, J. Hills and A. Sen (eds.), *Social security in developing countries*, Clarendon Press, Oxford.

Poulton, C., J. Kydd, S. Wiggins and A. Dorward (2006), "State intervention for food price stabilisation in Africa: Can it work?", *Food Policy*, Vol. 31, pp. 342-356.

Rapsomanikis, G., D. Hallam and P. Conforti (2004), "Market integration and price transmission in selected food and cash crop markets of developing countries: Review and applications", *Commodity Market Review, 2003-2004*, FAO, Rome.

Rapsomanikis, G. and A. Sarris (2008), "Market integration and uncertainty: the impact of domestic and international commodity price variability on rural household income and welfare in Ghana and Peru", *Journal of Development Studies*, Vol. 44, pp. 1354-1381.

Rosen, S. and S. Shapouri (2009), "Rising food prices intensify food insecurity in developing countries", *Amber Waves*, Vol. 6.

Rosen, S., S. Shapouri, K. Quanbeck and B. Meade (2008), *Food security assessment, 2007*, Economic Research Service, United States Department of Agriculture, Washington, DC.

Salant, S.W. (1983), "The vulnerability of price stabilization schemes to speculative attack", *The Journal of Political Economy*, Vol. 91, pp. 1-38.

Sarris, A. (2009), "Hedging cereal import price risks and institutions to assure import supplies", *International Agricultural Trade Research Consortium (IATRC) Meeting*, Fort Myers, Florida.

Sarris, A., L. Taylor and P.C. Abbott (1979), "Grain reserves, emergency relief, and food aid" in W.R. Cline (ed.), *Policy alternatives for a new international economic order*, Praeger Publishers, New York.

Sarris, A.H. (1998), *The evolving nature of international price instability in cereal markets*, FAO Commodity and Trade Division, FAO, Rome.

Sarris, A.H. (2000), "World Cereal price instability and a market based instrument for LDC Food Import Risk Management", *Food Policy*, Vol. 25, pp. 189-209.

Shapouri, S. and C. Christensen (2008), *Biofuels and higher global commodity prices: Impacts on food demand, commodity production, and trade in Sub-Saharan Africa*, Economic Research Service, USDA, Washington, DC.

Timmer, C. (2002), "Agriculture and economic development", in B. Gardener and G. Rausser (eds.), *Handbook of agricultural economics*, Elsevier Science, Amsterdam.

Timmer, C.P. (2008a), "Causes of high food prices", *ADB Economics Working Paper Series 128*, Asian Development Bank, Manila.

Timmer, C.P. (1989), "Food price policy: the rationale for government intervention", *Food Policy*, Vol. 14, pp. 17-27.

Timmer, C.P. (2008b), "Global commodity price rises and impacts on developing Asia", *Asian Development Outlook 2008 Update*, Asian Development Bank, Manila.

Timmer, C. (2000), "The macro dimensions of food security: economic growth, equitable distribution, and food price stability", *Food Policy*, Vol. 25, pp. 283-295.

Tollens, E.F. and C.L. Gilbert (2003), "Market liberalisation jeopardise export Quality? Cameroonian cocoa, 1988-2000", *Journal of African Economies*, Vol. 12, pp. 303-342.

Torero, M. (2009), "A system of virtual global reserves", in *Workshop on Institutions and Policies to Manage Global Market Risks and Price Spikes in Basic Food Commodities*, FAO, Rome.

Trostle, R. (2009), "Fluctuating food commodity prices: a complex issue with no easy answers", *Amber Waves*, Vol. 6, pp. 11-17.

Tschirley, D., J.J. Nijhoff, P. Arlindo, B. Mwinga, M.T. Weber, and T.S. Jayne (2004), "Anticipating and responding to drought emergencies in Southern Africa: Lessons from the 2002-2003 experience", *NEPAD Regional Conference on Successes in African Agriculture*, Nairobi, Kenya.

Tyers, R. and K. Anderson (1992), *Disarray in World Food Markets: A Quantitative Assessment*, Cambridge University Press, Cambridge.

Tyner, W. E., Serghini, H. and Ouraich, I. (2010), *Moroccan Agricultural Policy – A recent historical context and moving forward via the Maroc Plan Vert*, World Bank, Washington, DC.

UN High Level Task Force on the Global Food Security Crisis (UNHLTF) (2008), *Comprehensive Framework for Action*, United Nations, New York.

Valero-Gil, J.N. and M. Valero (2008), "The effects of rising food prices on poverty in Mexico", *Agricultural Economics*, Vol. 39, pp. 485-496.

Viatte, G., J.D. Graaf, M. Demeke, T. Takahatake and M.R.D. Arce (2009), *Responding to the food crisis: Synthesis of medium-term measures proposed in inter-agency assessments*, FAO, Rome.

Von Braun, J. (2008), "Food and financial crises: implications for agriculture and the poor", *Food Policy Report 20*, International Food Policy Research Institute, Washington, DC.

Von Braun, J. and M. Torero (2009), "Implementing physical and virtual food reserves to protect the poor and prevent market Failure", *IFPRI Policy Brief 10*, IFPRI, Washington, DC.

Wilcox, M.D. and P.C. Abbott (2004), "Market power and structural adjustment: The case of West African cocoa market liberalisation", presented at the American Agricultural Economics Association Annual Meeting, Denver, Colorado.

Williams, J.C. and B.D. Wright (1991), *Storage and commodity markets*, Cambridge University Press, Cambridge.

Wodon, Q. and H. Zaman (2008), "Rising food prices in sub-Saharan Africa: Poverty impact and policy responses", *Policy Research Working Paper 4738*, World Bank, Washington, DC.

World Bank (2007), "Agriculture for development", *World Development Report*, World Bank, Washington, DC.

World Bank (2008a), *Rising food and fuel prices – addressing the risks to future generations*, World Bank, Washington, DC.

World Bank (2008b), "Food price crisis in Africa", *World Bank Research Digest*, Vol. 3, pp. 1-2.

World Bank (2008c), *Global economic prospects 2009: Commodity markets at the crossroads*, World Bank, Washington, DC.

World Bank (2009), *Protecting progress: the challenge facing low-income countries in the global recession*, World Bank, Washington, DC.

World Food Program (2008a), *Recent food price developments in most vulnerable countries*, Issue 1, World Food Program, Rome.

World Food Program (2008b), *Recent food price developments in most vulnerable countries*, Issue 2, World Food Program, Rome.

World Food Program (2008c), *Summary of Price Impact Assessment Findings, Food Security Analysis Service*, World Food Program, Rome.

World Food Program (2009), *Trends in staple food prices in selected vulnerable countries*, Food Security Analysis Service, World Food Program, Rome.

Wright, B. (2001), "Storage and price stabilization", in B.G.A.G. Rausser (ed.), *Handbook of Agricultural Economics*, Hopkins University, Baltimore, MD.

Wright, B.D. (2009a), "Grain price gyrations: Recent research and its policy implications", *35th Annual James C. Snyder Lecture*, Purdue University, West Lafayette, IN.

Wright, B.D. (2009b), "International grain reserves and other instruments to address volatility in grain markets", *ARD-DEC Series*, World Bank, Washington, DC.

Wright, B.D. (2009b), "Speculators, storage, and the price of rice," *Giannini Foundation of Agricultural Economics*, Berkeley, CA, Giannini Foundation of Agricultural Economics, University of California, Berkeley.

Young, L. and Abbott, P. (2008), "Food aid donor allocation decisions after 1990", *Canadian Journal of Agricultural Economics*, Vol. 56, pp. 27-50.

Chapter 5

The Use of Input Subsidies in Low-Income Countries

by
Steve Wiggins and Jonathan Brooks

Input subsidies provide an operationally simple and politically attractive way of addressing multiple objectives. Economic objectives include stimulating production, offsetting high transport costs and input supply costs, making inputs affordable to farmers without credit, and allowing farmers to learn about the benefits of new inputs. In addition, they serve the social objective of transferring income to poor farmers. Input subsidies are an indirect, and relatively inefficient, way of addressing these objectives; they distort the allocation of resources, are prone to capture by vested interests, and often become fiscally costly. Innovative design features can mitigate some of these problems, but if input subsidies are deemed to have a short- to medium-term role it is important that their use does not crowd out spending on essential public goods, or compromise a long term approach of eliminating market failures – as opposed to offsetting them – and getting private markets working without subsidies.

5.1. Introduction

Interest in input subsidies, particularly those on fertiliser, has revived, notably in Africa. Yet orthodox economic advice suggests that subsidies on private goods should be avoided, since they distort resource allocation and are often costly and difficult to sustain without cutting spending on valuable public goods. They are also inefficient, and often inequitable, as a way of transferring resources. In the case of high income countries, OECD analysis suggests that less than half the value of an input subsidy translates into higher net incomes for farm households, with the majority of the transfer leaking to input suppliers or incurred as efficiency losses (OECD, 2001).

Why, then, the resurgent interest? A major concern is that whereas fertiliser use in developing countries has expanded rapidly over the past 40 years, uptake has remained particularly low in Africa (Heisey and Norton, 2007). At the same time, increased agricultural production to reduce poverty and improve food security has been stymied by the inability of many farmers to use the inputs and technology that are known to be effective (Duflo, Kremer and Robinson, 2008). In more detail, the argument, as applied to Africa, runs as follows:

- Food production per capita in Africa has grown too slowly, well behind rates seen in Asia and Latin America (Figure 5.1).

Figure 5.1. **Food production per capita since 1970**

1970 = 100

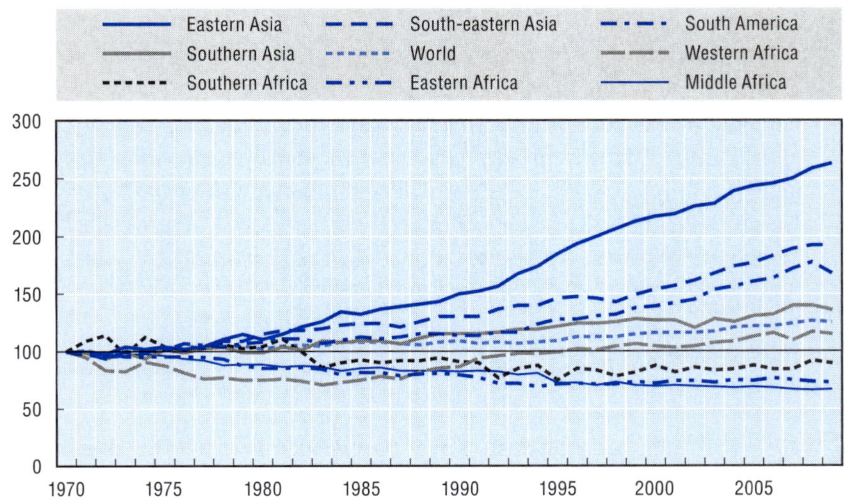

Source: FAOSTAT.

StatLink http://dx.doi.org/10.1787/888932552798

- This has resulted in rising imports of cereals and other staples, and more people who are hungry and undernourished (Figures 5.2 and 5.3).

Figure 5.2. **Cereal imports to Africa**

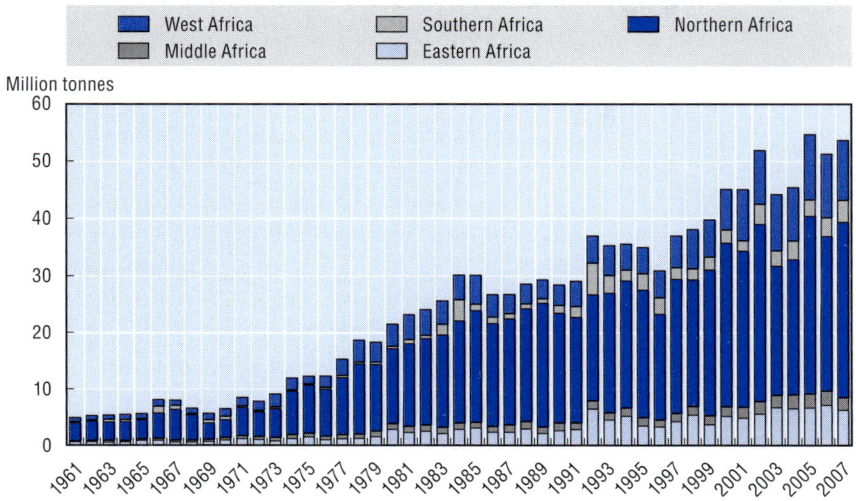

Source: FAOSTAT.

StatLink http://dx.doi.org/10.1787/888932552817

Figure 5.3. **Numbers of undernourished in Africa**

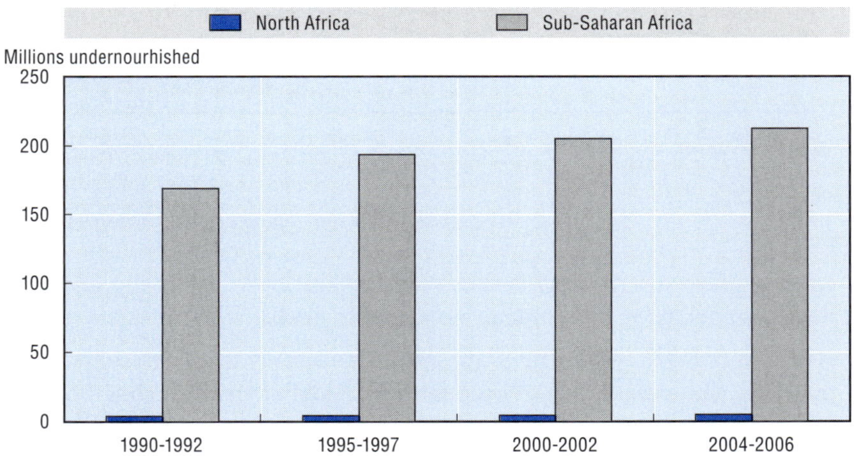

Source: FAO data on estimated numbers undernourished.

StatLink http://dx.doi.org/10.1787/888932552836

- Yields of staples per hectare have barely risen at all in the region (Figure 5.4), largely since farmers have not applied manufactured fertiliser in sufficient amounts (Figure 5.5) to take advantage of improved varieties.

Figure 5.4. **Staple food yields in Africa since early 1960s, three-year moving averages**

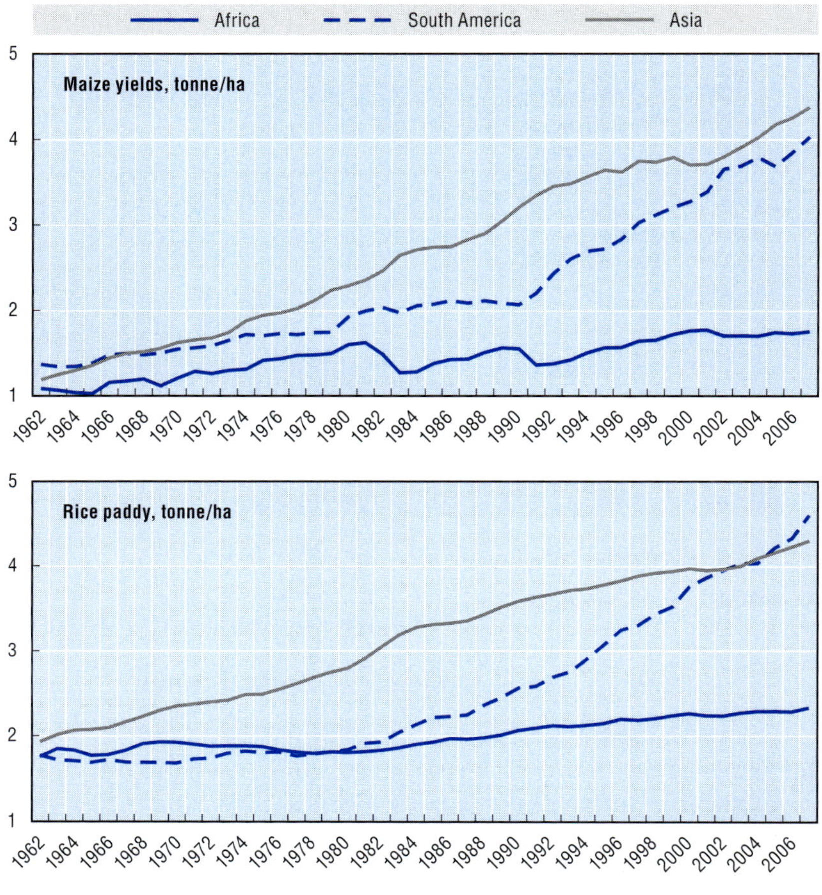

Source: FAOSTAT.

StatLink http://dx.doi.org/10.1787/888932552855

- Farmers have not done so because inputs have been too costly and they have been too poor, with little or no access to credit.
- Hence in order to resolve the impasse it is necessary to subsidise the costs of inputs, thereby creating a virtuous circle of higher yields, higher incomes, more food, less hunger and poverty.

The argument for subsidies is attractive on several grounds.

- "… the siren call of subsidies continues to be hard to resist; they are politically attractive, seem easy to implement, and the problems they are intended to address remain compelling at both the national and international levels" (Crawford, Jayne and Kelly, 2008).

Figure 5.5. **Fertiliser application rates, kg nutrients/ha arable land, 2006/08 average**

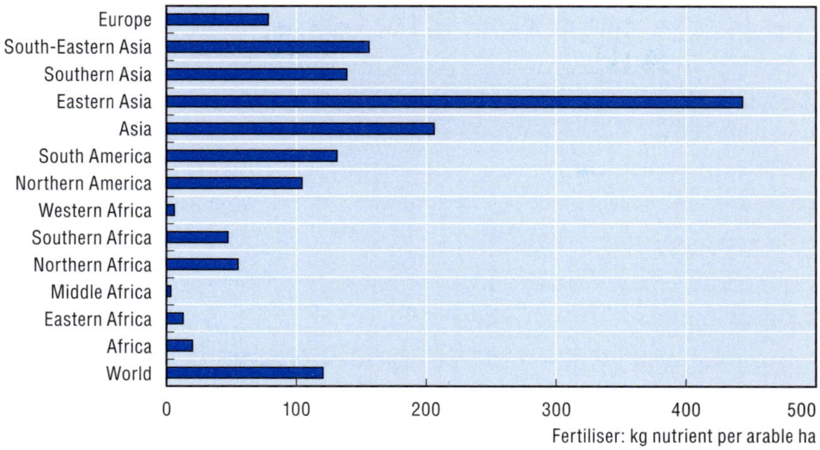

Source: FAOSTAT.

StatLink http://dx.doi.org/10.1787/888932552874

- Subsidies look as though they provide a ready solution to otherwise difficult problems of developing input markets and associated financial services to small farmers. While other ways of overcoming such problems are complicated, with success uncertain, a subsidy is a relatively straightforward measure to implement.

Politically, subsidies are a highly visible gesture to rural voters, as well as potentially also being an instrument of patronage. To some, moreover, the sight of the state taking direct action to overcome the failings of markets, often regarded with suspicion in the first place, is welcome. Moreover, in parts of the developing world orthodox advice to avoid subsidies is treated with suspicion, since it usually comes from people in countries where farmers are subsidised.

Yet perhaps the greatest attraction lies in the apparent simplicity of a single measure, a subsidy, to meet a wide range of objectives: economic, social and political. It is thus necessary to unpick the different justifications that can be offered for subsidies. Section 5.2 therefore discusses the multiple objectives of input subsidy programmes. The remainder of the report is then structured as follows: Section 5.3 considers some of the disadvantages and dangers associated with their adoption, while Section 5.4 examines the experience of input subsidy programmes in three countries: Malawi, Sri Lanka and India. Given that the effectiveness of input subsidies depends to a large extent on how they are implemented, Section 5.5 considers factors relevant to the design of input subsidy programmes. Finally, Section 5.6 presents some policy conclusions.

5.2. Unpicking the arguments: Objectives of input subsidy programmes

The broad argument that input subsidies can lead to higher incomes, reduced poverty and improved food security is based on specific claims with respect to a range of underlying objectives. Most of these underlying objectives have either an economic efficiency rationale (*i.e.* reflect a market failure of some kind), or are concerned with reallocating income to a particular constituency (a social objective). The main economic objectives are as follows.

- *Stimulate agricultural production.*
- *Compensate for high costs of transport* from port or factory to farms that raise costs of inputs.
- *Improve soil quality* and combat soil degradation (in the case of fertiliser).
- *Offset high costs of supplying inputs* when markets have low volumes and economies of scale in logistics cannot be achieved.
- *Make inputs affordable* to farmers who cannot buy them, owing to poverty, lack of access to credit, and inability to insure against crop losses.
- *Learning* – to allow farmers to try novel inputs and become familiar with their advantages.

In addition, there is a social objective.

- *Social protection* – to transfer income to farmers who are poor, live in remote disadvantaged areas, or both.

In practice, it may be difficult to disentangle these explicit objectives, particularly the social one, from the implicit motive of political patronage: to win favour with voters and reward supporters.

Raising the level of output (objective 1) is not in itself an efficiency issue. However, this objective reflects the notion that output may be less than optimal because of underlying market failures, for example the sub-optimal use of inputs, and the possibility that higher output could lead to external economies of scale (a "thickening of markets"). The short- and long-term benefits of using input subsidies need to be compared with the costs of tackling those market failures directly.

The same goes for objectives 2 and 3. The optimal way of offsetting high transport costs is to invest in infrastructure. Similarly, while applying more fertiliser can raise soil fertility and prevent soil degradation, there are other ways of doing this that may be more effective and economical, such as terracing to prevent soil erosion. Moreover, too much application of fertiliser can lead to environmental degradation.

Arguments four and five are central to current debates: these aims are about correcting market failures affecting input supply, finance and insurance. One set of failures (objective 4) arises in the supply chain. When potential input dealers know too little about the demand for inputs, and suspect that demand may be low, they will not stock them. What little stock they may carry then has a high mark-up to cover both the risk of not being sold as well as high unit costs for transport and storage of small lots, since scale economies are not achieved. Another set of failures (objective 5) affects farmers. They lack the cash to buy inputs early in the crop season and cannot obtain credit. Banks or input dealers will not offer credit if they do not know enough about the competence and character of farmers seeking loans, or will only do so if they can get collateral and character references – requirements that many small farmers cannot meet. Farmers, moreover, may be reluctant to accept the risk of credit in any case, since they would be unable to repay the loan if the harvest fails. Formal insurance policies are usually absent in rural Africa, since would-be insurers face similar problems to bankers: the underlying risks are difficult to calculate, the character of farmers is unknown. Offering them insurance would be foolhardy without this information. However, there may be scope to create viable insurance and micro-credit alternatives to input subsidies.

If market failures are severe, farmers could become locked into low levels of productivity, even when the technology and economic opportunity exist, since they cannot access and afford the seeds and inputs to take advantage; and thus they remain trapped in poverty, too poor to work themselves out of this condition (Dorward *et al.*, 2004; Duclos and O'Connell, 2008). If this applies to many farmers, as it may in some countries where the majority of rural households are poor, then a household poverty trap becomes a major drag on national economic growth as well. It is not then surprising that there have been calls for governments to intervene to correct the failures, if necessary by subsidising costs; and if necessary by providing inputs directly to farmers. A lively current debate in Africa turns on how widespread and severe are these rural market failures; and whether there are other ways of remedying them than input subsidies.

Reason number six, on learning, is another form of market failure. Farmer demand for improved inputs may be low simply because they have too little experience of their advantages. There is a strong case for a subsidy in such cases, but since farmers can try out inputs and assess their advantages within a couple of seasons, a subsidy on these grounds would be short-lived. Moreover, since farmers tend to try out new ideas on limited areas, the subsidy need only cover a small amount of seed, fertiliser, chemical, etc., per farmer; there is no need to a blanket subsidy in such cases. It is thus not surprising that a common alternative to a subsidy for learning is to distribute, for free, starter packs with improved seed and fertiliser sufficient to plant an acre or less.

Finally, argument seven, the use of input subsidies to transfer income to poor farmers or those disadvantaged by location needs to be assessed against the effectiveness and economy of doing the same by direct payments,[1] food aid distribution, or employment programmes paid in cash or kind.

For all of the market failure (efficiency) reasons commonly cited for subsidising inputs, the theoretically optimal response is to correct those market failures directly. Input subsidies, by tackling the problem indirectly, risk inducing unintended secondary effects. Similarly, input subsidies are not the optimal way of transferring incomes to a target constituency. The arguments therefore come down to the relative cost-effectiveness of different instruments and practicalities of implementation. In this respect, it is important to consider the disadvantages and drawbacks associated with input subsidies.

5.3. Disadvantages and dangers of input subsidies

Arguments against subsidies include the following.

- Subsidies may be ineffective in raising use of inputs and increasing yields. It is not always the case that the volume of inputs applied is sensitive to price. Studies in Sri Lanka (see below), for example, report low elasticities of fertiliser application with respect to its own price: instead the volume of fertiliser applied corresponds more closely to the area under irrigated rice and to the price of rice. The corollary in these cases is that much of the subsidised fertiliser merely displaces fertiliser that would have been bought without the subsidy.
- Heavy subsidies on inputs potentially distort the relative costs of factors, leading to inefficient allocation of inputs, with the subsidised inputs substituted for other factors. This applies particularly where inputs are substitutes, rather than cases where they are complementary. The most often cited case for agriculture is that of farm machinery, where capital grants and tax exemptions for tractors and harvesters lead to farmers using machinery to displace day labourers – in places where there are many landless looking for work. Fertiliser is typically thought of as a complementary input, but there may be cases in which it is a substitute.[2]
- Subsidies intended to benefit specified groups of farmers, or to stimulate particular crops, may be less effective than intended as leakages occur. For example:
  - When farm profits rise, landlords may be able to raise land rents and thus effectively appropriate the value of the subsidy. The degree of this leakage depends on the extent to which farmers rent rather than own land (or have otherwise secured land rights), and the extent to which the price of land is bid up. The nature of these leakages is assessed using the DEVPEM model (OECD, 2011).

- When subsidy programmes allow discretion to local officials and field workers in allocating subsidies inputs, there is the danger they will use their power to extract bribes.
- The same local discretion may be used to divert subsidised inputs from intended beneficiaries to others, such as local elites and political supporters. In some cases this arises since field workers have different priorities to policy-makers. For example, in Malawi some field staff reportedly prefer to allocate subsidy vouchers to farmers they consider most likely to make good use of the input, rather than those who cannot afford fertiliser at commercial prices (Dorward and Chirwa, 2011).
- When subsidised inputs are intended for use on a particular crop – often food, they may be switched to higher value cash crops. For example, some of the subsidised fertiliser in Sri Lanka intended for rice is reportedly diverted to vegetable production. This may be a rational response from farmers and raise their economic welfare, but it subverts the intended objective, which in the case of Sri Lanka is greater self-sufficiency in rice.

- Subsidy programmes may be implemented in ways that repress the development of private supply of inputs by delivering inputs through state agencies and bypassing nascent local input dealers.
- In the absence of effective targeting, input subsidies are likely to benefit larger farmers, whose use of inputs is highest, and may fail to reach the poorest and most vulnerable households. These distributional effects are also explored using the DEVPEM model (OECD, 2011).
- When subsidised inputs dominate the supply of a particular input, then subsidies may become closely linked to government budget cycles, or to electoral cycles with pronounced swings in availability so that supplies, whether or not they are subsidised, may not be regular, reliable and timely.
- All this said, often the main objection to subsidies is their high cost. In the three cases outlined below of India, Malawi and Sri Lanka, input subsidies have taken up 10%, 15% or more of the total government budget – sometimes more than what the government spends on education. When state agencies are used to distribute subsidised inputs, there may be little incentive for them to economise on logistics.
- What is more, the cost of subsidies can rise if:
  - the subsidy is effective in encouraging greater use of inputs;
  - policy makers are tempted to use the subsidy as a way to stabilise input prices, so that when the cost of inputs rises, the subsidy is used to maintain the same nominal cost to producers;

  - subsidised inputs are smuggled across borders to neighbouring countries where inputs are more expensive (a problem if the subsidy is not rationed or otherwise limited); and, if
  - political competition for rural votes leads parties to make election pledges to increase subsidies.
- Once in place, subsidies can be difficult to remove. They can be seen as a political signal of support to farmers, around which farmers sometimes form electorally powerful lobbies for their continuation.

The often high costs of subsidy programmes need to be set against the benefits they create and counted in terms of the missed opportunities to use public funds for other purposes. There cannot be a general judgment on the balance between potential benefits and disadvantages of subsidies; so much depends on particular circumstances and the design of the programme. The next section reviews the experiences of input subsidies in three countries where programmes were implemented to make up for the absence of a well-functioning domestic market for inputs.

5.4. Experiences of input subsidies

Malawi's fertiliser subsidies[3]

Fertiliser subsidies in Malawi date back to the mid-1970s, but were suspended in the early 1990s under liberalisation. Targeted starter packs of seed and fertiliser sufficient to plant one-tenth of a hectare were introduced in 1998/99, intended to allow small farmers to learn of the benefits of the package, to increase food production, and improve their incomes. These were effective in raising output of the main staple, maize, but not sufficient as the country experienced poor harvests in 2001, 2002, 2004 and 2005. In response, the government brought back universal subsidies on fertiliser in 2005/06, much to the consternation of some donors.

In 2006/07 two million seed and three million fertiliser vouchers were distributed to targeted households. The vouchers allowed recipients to buy two 50 kg bags of fertiliser at what was then 28% of full cost. In total, 175 000 tonnes of fertiliser and 4 500 tonnes of improved maize seed were distributed at a cost of USD 91 million. By 2008/09, 182 300 tonnes of fertiliser for maize were made available with vouchers planned for 1.5 million households.

Judging the impact of the Malawi programme is complicated. Supporters point out that maize production has increased remarkably since the introduction of the subsidies (Figure 5.6), with harvests exceeding the estimated national requirement of around 2.4 million tonnes. While some of the production increase in the last four years may be the result of favourable rains, it would be hard to deny the impact of the subsidies.

Figure 5.6. **Malawi: Maize production, 1990 to 2009**

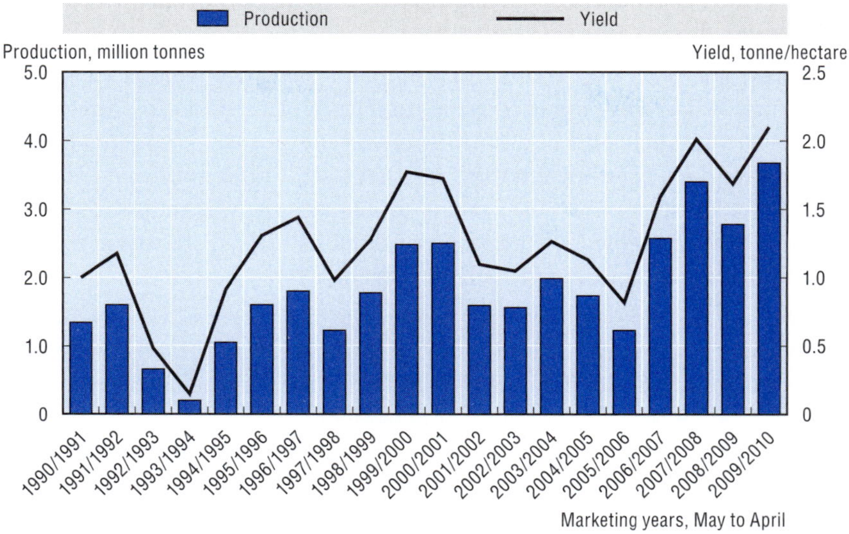

Source: USDA data.

StatLink http://dx.doi.org/10.1787/888932552893

The programme, however, has not been without difficulties. Targeting has been imperfect, and some of the vouchers have been distributed to political supporters. Although vouchers are used, fertiliser has been distributed by a state company in recent years, marginalising small private dealers in rural areas. Politically, parties have promised to increase the programme, recognising it as a vote winner, regardless of the economic merits of expansion. In 2005/06, 8% of the expenditures of farm households were on subsidised inputs. The overall cost of the programme has been rising as well: to more than USD 200 million during 2008/09, representing 16% of the total government budget. This is partly due to expansion of the programme, but more the rise in world fertiliser prices, which led to an increase in the unit value of the subsidy.

Measuring impacts more precisely is complicated by questions about what might have happened without the subsidies – not all the fertiliser distributed was additional, with estimates of 20%-30% displacement of commercial sales in some years – as well as by consideration of second-round effects on maize prices, rural employment and wages, when dealing with a crop that is so central to the economy. Since transport costs from Indian Ocean ports to Malawi are substantial, there is wide band between import and export parity prices for maize in Malawi: unless harvests fail badly, the domestic price for maize is largely determined by the size of the domestic harvest. Hence higher maize production has the potential to bring down domestic prices, to the benefit of low-income consumers, including very small farmers, who are usually net buyers of grain. Maize prices in the last few years, however, have been surprisingly high given the bumper harvests, even when taking some limited exports into account.[4]

There are, however, reports of more maize being available in villages, of rising rural wage rates, and of reduced rural poverty. Fertiliser importers have benefitted from increased public procurement. But for local retailing dealers, the impacts have been mixed: they may have sold more improved seed, but they have lost sales of fertiliser owing to displacement by subsidised sales. This was not inevitable, but the result of administering distribution through public channels rather than the private local dealers.

Overall, the judgment on the Malawi experience turns on how effective the fertiliser subsidy has been in raising the harvest and the second-round effects this has had in a low-income economy where staple food production has significant effects on both rural employment and the cost of living. In this regard, the particular circumstances of Malawi need to be taken into consideration: high levels of rural poverty; few sources of rural employment other than agriculture, and not many jobs in urban areas either; weakly developed rural markets for inputs, credit and insurance; low incomes throughout the economy so that staple food makes up a significant share of the cost of living; and remote location from world markets so that domestic harvests largely determine domestic price of maize.

Sri Lanka's subsidies on fertiliser for paddy farmers

Sri Lanka has subsidised the cost of fertiliser, with a short interruption in the early 1990s, since 1962:

> "... with the intention of encouraging the use of fertilisers and off-setting the effects of low crop prices and high costs of production" (Tibbotuwawa, 2010).

The programme since 2005 ensures that rice paddy farmers can obtain fertiliser at a fixed price, the government paying a subsidy to importers to cover the difference between the fixed price and the imported cost. In 2010, the price of a 50 kg bag of fertiliser was reported as just USD 3.07: a 93% subsidy on the cost of urea[5] (Tibbotuwawa, 2010). Total cost to Sri Lanka has risen to reach USD 570 million, or 3% of gross domestic product.

How effective is this very heavy subsidy? For production, the question is the extent to which the subsidy encourages additional use of fertiliser and does not simply replace what would have been bought commercially. Some studies show a low responsiveness of fertiliser use on paddy rice to the price of fertiliser. Kikuchi and Masao (1990) estimated that removing the fertiliser subsidy would reduce rice yields by only 1%-2%, since nitrogen demand was relatively inelastic with respect to price. This conclusion was supported by Ekanayake (2006), who found low elasticity of fertiliser prices on consumption for the three main fertilisers. In his analysis the price of rice was more important in determining fertiliser use than the own price of fertiliser.

The subsidy is targeted to rice farmers – although there are reports of leakage to vegetable growers – with no restrictions on the amount sold with subsidy. The World Bank (2003) estimates that about 51% of the total rice area is cultivated by households in the top 40% of the rural expenditure quintiles while only 25% is cultivated by the households in the lowest two quintiles. Hence, as a way to transfer income to the poor, the subsidy appears a blunt instrument.

India's subsidies on fertiliser, irrigation water and rural electricity

India originally introduced subsidies in the 1960s to support the green revolution, with major spending to keep down the costs of fertiliser, irrigation water from public systems, and rural electricity – that drove many of the private irrigation pumps attached to wells. In addition India nationalised the main banks and directed them to provide credit to farmers at concessional interest rates.

While studies (Dorward *et al.*, 2004; Smith and Urey, 2002) suggest that during the early phases of the green revolution payment of subsidies on inputs contributed to rapid expansion of production of cereals and thereby to poverty reduction, subsequently it is less clear that the subsidies have continued to do so.

Input subsidies also became a major feature of policy and were valuable to farmers faced with declining output prices in the 1980s. They were not, however, key determinants of technology adoption and became damaging when they crowded out capital investment in research, infrastructure and human capital as fiscal constraints began to bite under structural adjustment reforms. They may also contribute to environmental degradation and, to the extent that wealthier farmers and regions largely captured them, they had little direct influence on poverty (Smith and Urey, 2002).

Meanwhile since the early 1980s the costs of the subsidies have increased notably (Figure 5.7), rising to between 15% and 25% of the government budget, more than is spent on education, in a country where adult literacy is only 61%.

Since the mid-1990s, agricultural production increases have slowed, apparently for lack of investment in physical infrastructure, research and extension. The fear is that the cost of the subsidies has crowded-out other public investments, as Figure 5.8 would suggest.

Figure 5.7. **India: Cost of agricultural subsidies, 1980 to 2002**

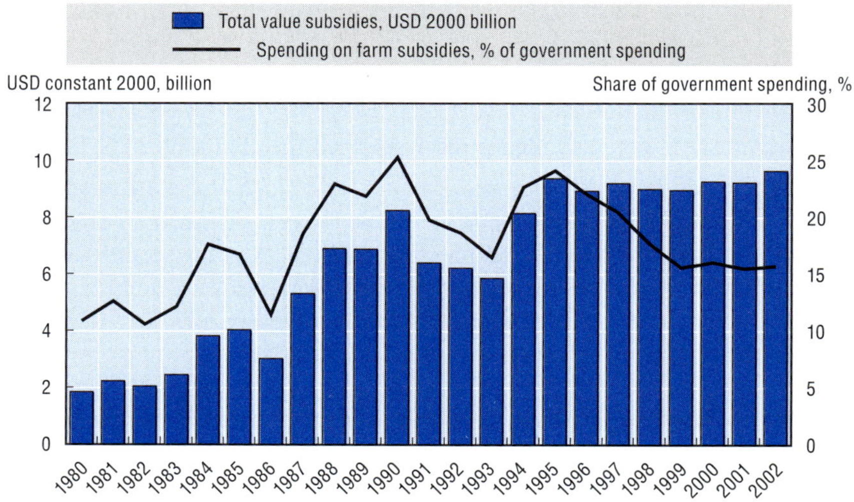

Source: Mullen *et al.*, 2005.

StatLink http://dx.doi.org/10.1787/888932552912

Figure 5.8. **India: Spending on subsidies compared to public goods**

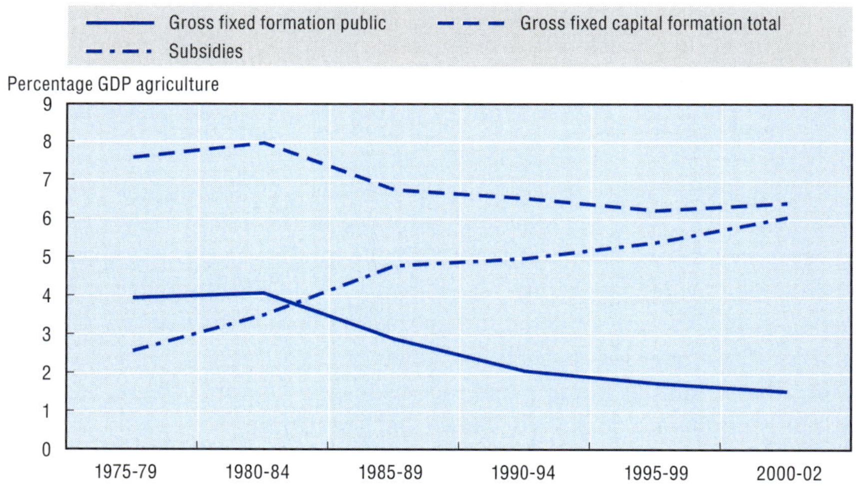

Source: Chand and Kumar, 2004.

StatLink http://dx.doi.org/10.1787/888932552931

Lessons from these cases

Evidence from case studies of India, Malawi and Sri Lanka suggests that subsidies have had an impact over the short- to medium-term, promoting input use, raising output and thus reducing poverty. The programmes have been costly, although the absence of counter-factuals makes it difficult to evaluate whether the same benefits could have been achieved at a lower cost with alternative instruments. It is also possible that, because of high budgetary costs, the pursuit of other objectives, for example in the areas of health and education, has been compromised.

A major question mark hangs over whether the benefits of the programmes have been enduring, in the sense that they have led to a sustained increase in incomes that would survive removal of the subsidy. Experience from India is that there were early returns during the Green Revolution, but the subsidies became increasingly ineffective as they were not complemented by deeper investments to improve agricultural productivity and strengthen the rural economy. Hence there was relatively weak progress in facilitating the agricultural transformation and raising rural incomes. In general, for subsidies to have had any long-term effect, they require complementary investments to make input use profitable, for example in rural roads, agricultural research and extension, and, in some cases, irrigation. Indeed, the extent of adoption of high-yielding varieties and use of irrigation may have more influence on the amount of fertiliser used than the price of fertiliser.

There is also evidence that the benefits of input subsidies are higher in the early stages of provision, as farmers increase their use of external inputs from a low base. They are markedly lower once a certain level of use has been achieved, agricultural production is greater and markets have become wider. Furthermore, the tendency is for costs to rise, for the subsidies to increasingly displace government spending in other areas, and for them to become a source of income transfer from which the government has difficulty extracting itself. Hence they can pass from being a help to becoming a hindrance to agricultural development.

The effectiveness of input subsidies will depend on specific market conditions and the way in which the subsidy programme is implemented. Evidence from Sri Lanka suggests that fertiliser use might not be sensitive to price (in which case the subsidy simply replaces commercial sales). An open-ended subsidy also favours larger producers, making it a poor instrument for tackling poverty. Effectiveness may also be constrained by design features. For example, the state's distribution of vouchers in Malawi has led to a diminished role for private dealers. Issues relating to the design and operation of input subsidy programmes are taken up in the next section.

5.5. Design of subsidy programmes

Much of the effectiveness of subsidies depends on how programmes are designed and operated. Considerations and options that arise include:

- Targeting: Should the subsidies be offered to all users, or targeted to specified farmer either on efficiency or social policy grounds, or for a combination of both? Possible criteria for targeting are farm size, location, via means-testing for poverty, production of particular crops. Targeting can make subsidies more likely to achieve their goals economically, but may be administratively costly and prone to leakages.

- Similar to targeting is the imposition of conditions on receipt of subsidy, such as use on particular crops, application according to specified techniques, or conditional on selling a set quantity of output to the state. These may improve the focus of the subsidy on the intended aims, but raise the costs of administration and monitoring for compliance.
- Rationing: the subsidy may apply to all of the input, or to a limited quantity for each farmer. This can thus become a way of targeting: if farmers are entitled to just a few bags of seed or fertiliser per farm, in effect the subsidy has much greater incidence on smaller and presumably poorer farmers, compared to larger farmers.
- The form in which the level of subsidy is specified. Subsidies may be computed as a percentage of cost, may be a fixed amount per unit volume of subsidy, or may hold the price of the input at a given level, that may be indexed to inflation or not. In the last case, holding the price constant risks seeing the subsidy bill rise when input prices rise, and it reduces the predictability of public spending.
- Point of provision of subsidy: Should it be paid to importers, wholesalers or retail dealers, or given to farmers in the form of vouchers? The higher up the chain, usually the fewer payments have to be made, which implies administrative savings, but with the danger that the subsidy will not be fully passed on to farmers. Providing farmers with vouchers ensures that the subsidy goes to the beneficiaries, but reaching hundreds of thousands of small farmers can be a major administrative exercise that takes up much of the time of extension staff.
- Distribution of subsidised fertiliser: Should the subsidy be supplied by private dealers, by co-operatives, or distributed by a state agency. Using state agencies can make it easier to target beneficiary farmers, areas or crops; but may inhibit the development of private marketing. State agencies may also have little incentive to reduce logistical costs.
- Duration of the subsidies: They may be limited in time, particularly when they aim to promote learning of use of inputs; or they may have no time limit.

Much depends on what the subsidy is intended to do. Table 5.1 draws the implications of different aims for programme design. Although the various motives can be broken down as shown, in practice more than one aim may apply. There is a risk, however, when a single policy is justified by more than one aim. While a single aim may allow the policy instrument to be designed

Table 5.1. **Different aims for input subsidies, different implications**

Aim	Implications for subsidy	Exit strategy	Alternatives and complements
Stimulate agricultural production	Target to farmers who make most use of inputs, perhaps those considered more competent, or those with better land and water	Only when a particular level of production has been reached, but could be unending	Improve marketing Invest in public goods for agricultural development, especially roads, irrigation and drainage, research and extension
Learning new technology	General subsidy; or target to non-users Only need to subsidise small amount of fertiliser: benefits should be visible on half hectare plot	Use until farmers see benefits from fertiliser: could be for as little as one or two seasons	Grant: distribution of trial packs of seed and fertiliser, sufficient for 0.1 ha or 1 acre. Permit fertiliser to be sold in small units: 5 kg or less
Improve soil quality	General subsidy	May be unending if fertiliser necessary to maintain soil quality and contain degradation	Soil erosion control through physical works and vegetation management Crop management practices to improve structure of topsoil, retain and recycle nutrients
Compensate for high transport costs	General subsidy, but may vary according to remoteness – potentially leading to pan-territorial pricing	When transport costs have fallen sufficiently	Invest in roads, ports, storage Encourage competition in transport, end cartels
Compensate for market too thin to achieve scale economies	General subsidy since all farmers suffer from market failure	Use until market works with lower transaction costs having reached scale economies and overcome information failures When economies of scale in transport and distribution apply, then opportunity for easy exit: as volumes traded rise and logistical costs fall, sub can be progressively withdrawn while marked price to farmers remains constant. Requires careful monitoring by government to time reduction and eventual withdrawal of subsidy	Foster institutional innovations to reduce transactions costs Dealer training, public underwriting of trade credits May require additional public investments in roads, ports, storage
Compensate for farmers too poor to afford inputs	Target subsidy to poor farmers	Use until poverty reduced amongst farmers	Stimulate financial services, including micro-finance and micro-insurance to allow poor to get inputs on credit Poverty reduction measures, including social protection
Social equity	Target subsidies to those who need help, by social groups, areas, or both	Exit when disadvantaged reach a threshold income level	Poverty reduction programmes Social protection and safety nets for disadvantaged, including: Cash transfers Pensions Employment programmes

specifically for that purpose, when there are several objectives the instrument tends to become the lowest common denominator that will serve all.

That would not matter if all aims were equally important and all achieved at the same time; but that is not the case. A policy that was originally introduced to meet several objectives may remain largely unchanged, even when most objectives have been met and the remaining aims indicate that policy needs corresponding adjustment. Programmes set up with multiple aims are an easier target for lobby groups: those interested in the continuation of the policy can usually find some reason or other from the original list to argue for this. Multiple objectives can make it more difficult to evaluate the programme with precision.

During the 2000s much has been said in favour of "smart" subsidies (World Bank, 2007), that is programmes where subsidies are:

- *Targeted* to those farmers for whom inputs may be otherwise unaffordable – for example, poor farmers, farmers in remote areas.
- *Work with the market* to help develop commercially viable supply chains – for example, by giving farmers vouchers that they can redeem from input dealers.
- *Limited in time*, until the market failures that justified the subsidy have been overcome.

These look like sound principles, although it is easy to see why they may not be followed. Administratively, targeting may be difficult and costly; and, if vouchers are distributed, there may be the need for complementary measures to ensure local dealers stock inputs – by providing, perhaps, trader credit. Politically there may well be pressure to make the benefits universal; apart from anything else, targeting can be socially and politically divisive if the criterion for access is a continuous and difficult to establish variable such as income, there may be distrust of the local dealers, and there will almost certainly be pressure to sustain the programme.

There is a close second-best to targeting: rationing, for example by limiting the volume of seed or fertiliser on which a farmer can receive a subsidy. This overcomes most of the administrative and political objections, while providing proportionately greater benefits of the subsidy to smaller and poorer farmers. The main objection to this applies if the subsidy is designed mainly to increase input use for higher production and it is argued that small farmers cannot make good use of the inputs.

Indexation of the subsidy may be one way of ensuring that the subsidy is time-bound and leads to the replacement of state action with private markets. When underdeveloped transport, storage systems mean that the farm-gate cost of inputs is high, a subsidy may allow use of inputs. If enough farmers adopt and use the inputs, then some of these costs may fall with economies of scale in transport, storage and distribution. Costs may be further reduced if there is sufficient spending on roads and ports with development. Hence it may be possible for low-income countries with the characteristics described above to adopt a medium-term strategy that sees a subsidy set with respect to a price indexed to inflation that would allow the subsidy to decline as logistical costs come down. The subsidy could thus become self-limiting in time, with input users not feeling the effects of the taper.

For this to happen, however, there has to be public investment in transport and fertiliser distribution needs to be in the hands of private firms that have the incentives to reduce logistical costs. Is this wishful thinking? The Kenyan experience with liberalised fertiliser markets (Box 5.1) suggests that these costs can be reduced substantially.

Box 5.1. **Kenya's experience of liberalised markets**

Fertiliser supply was liberalised in the early 1990s, when retail price controls, import licensing quotas, and foreign exchange controls were eliminated. At the same time, donor fertiliser deliveries were phased out. Private response to liberalisation has been good.

The network of dealers has expanded, to reach 500 wholesalers and 7 000 retailers, so that the average distance from farm to dealer fell from more than 8 km to just over 4 km between 1997 and 2004.

Improvements in logistics mean that the real cost of moving fertiliser from Mombasa to farms up country was cut by around 40% in real terms in the 1990s. As a result, use of fertiliser has increased.

Figure 5.9. **Price of fertiliser at the port of Mombasa and in Nakuru, centre of main area of commercial maize growing**

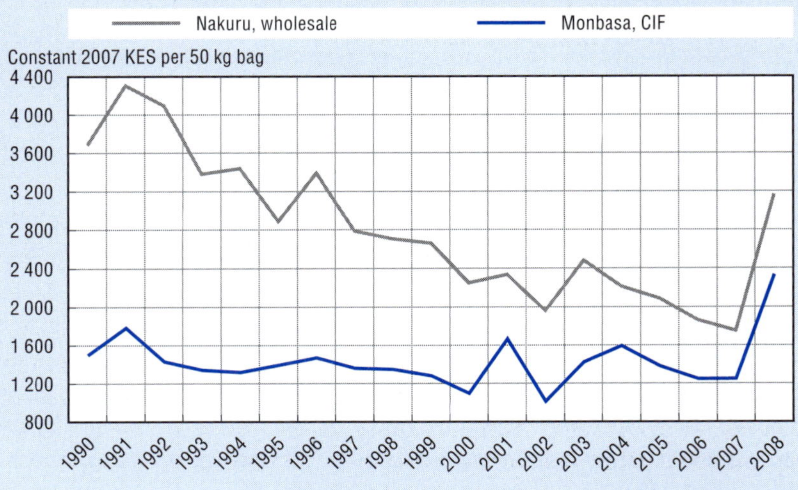

StatLink http://dx.doi.org/10.1787/888932552266

Box 5.1. **Kenya's experience of liberalised markets** *(cont.)*

Figure 5.10. **Fertiliser consumption and imports in Kenya, 1990 onwards**

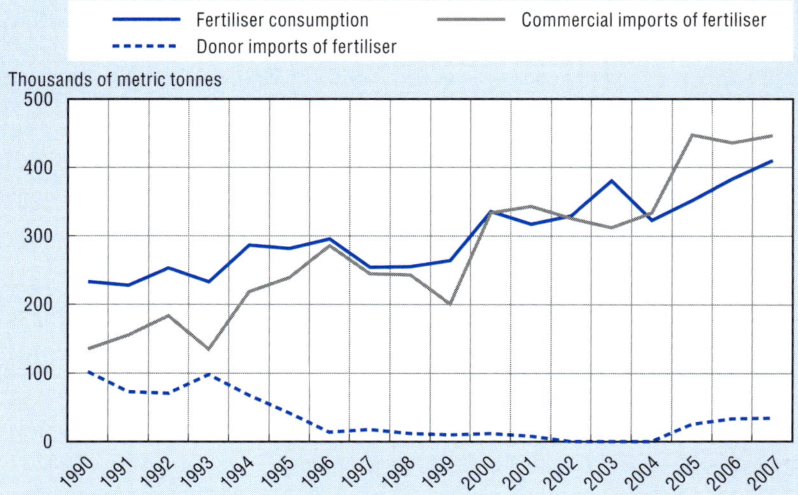

Source: Ariga and Jayne (2009).

StatLink http://dx.doi.org/10.1787/888932552285

Especially noteworthy is that increased use of fertiliser has been common amongst small farmers, who make up most of those farming in Kenya. In 1995/96, 43% of smallholders used fertiliser, while by 2006/07 this ratio was 70%. The amount of fertiliser applied per hectare has risen, reaching 190 kg/ha on maize, a level comparable with other parts of the developing world and well above typical levels seen in other parts of Africa.

Increased fertiliser use has led to higher yields. Maize yields on small farms are up, from an average of 1.48 t/ha in 1997 to 2.1 t/ha in 2007. These yields may not seem so high, but much of the maize on small farms in Kenya is intercropped with other plants, such as vegetables. Hence these yields measure only part of the production of many fields.

While some of the fertiliser has been supplied under contract farming for sugar cane, and under schemes that integrate marketing, credit and inputs for tea and cotton, much of the fertiliser has been supplied in cash deals through private input dealers.

Not all is well: the smallest farms and poor farmers still find it hard to buy fertiliser for lack of credit. In response, the government announced in 2007 a programme to subsidise fertiliser and maize seed sufficient for one acre, aiming to reach 2.5 million small farmers.

5.6. Conclusions

Input subsidies need to be contemplated with caution, with a clear consideration of the costs and benefits compared with conventional best practice of addressing market failures directly and using social policies to address social objectives with respect to poverty and food insecurity.

The technology transfer ("learning") argument is one that corresponds to best practice advice because of its public good aspect. However, this calls for a relatively small time-bound subsidy. What is being proposed, particularly in Africa, is intervention on a much larger scale, and for a longer duration, than warranted by this objective. In general, input subsidy programmes are being developed to address the specific circumstances of countries at a low level of economic development, where there are:

- extensive and severe market failures with resulting poverty traps, so that any increased production has directly reduces poverty and raises human capability; and
- locations where food markets are naturally protected by distance from other markets, so that local prices will be driven down by increases in local production.

In these cases, a time-bound input subsidy may provide an alternative to failing markets, leading to more use of the input, with higher production that then raises the incomes of farmers, provides more work for agricultural labourers, and reduces the cost of food, allowing those on the breadline to consume more and become more productive. The subsidy then could become an element in breaking through limits to growth and shifting both the agricultural and national economies to a path of faster growth.

In order to achieve these benefits, there will be a need for complementary spending on public goods. For agriculture, these usually consist of rural roads, agricultural research and extension, education, primary health care, and clean water. A further requirement is the establishment, or maintenance, of an adequate rural investment climate – including peace and political stability, a competitive exchange rate, low inflation, modest interest rates, fair taxes, and in general policies that allow investors to have reasonable confidence that they can carry out their business without undue hindrance now and in the near future.

In the longer run, however, it is much less easy to justify input subsidies. Indeed, they frequently become a millstone that eats into the funds for public goods and thereby undermines growth and development. Politically, once in place subsidies are difficult to remove.

If these difficulties are to be surmounted, there are important lessons that need to be absorbed. These lessons are reflected in the principles for "smart" subsidies espoused by the World Bank and others. Subsidies should be targeted as effectively as possible, work with the market (for example, by giving vouchers to farmers rather than have the state distribute subsidised fertiliser) and be time-bound (with indexation a possible way of avoiding an indefinite government commitment).

Above all, the rationale for using a subsidy needs to be kept clear. Despite the attraction, there is a fundamental difficulty in using a single instrument to address multiple market failure (*i.e.* long-term development) and social objectives. With respect to market failures, there needs to be an exit strategy. This in turn means that a system of social protection needs to be put in place, and be fully operational once the subsidy is withdrawn. The conflict between economic and social objectives has become a clear problem in India and Sri Lanka, where programmes have outlived their original rationale and become a budgetary millstone.

Finally, as far as possible, the use of input subsidies should not crowd out spending in other important areas, or compromise a long-term approach of eliminating market failures – as opposed to offsetting them – and thereby getting private markets working.

Notes

1. Cash transfers may be conditional on the clients sending children to school, or on mothers attending primary health care clinics with infants.
2. For example, women farmers in Swaziland have been seen to substitute fertiliser for weeding: they are chronically short of labour, but have remittances from their miner husbands in South Africa, and hence it makes sense to push up yields with fertiliser even as yields are depressed by weeds.
3. Main sources: FAC 2008, 2009; Dorward and Chirwa, 2011.
4. Dorward and Chirwa (2011) discuss why this has happened, breaking with a previous pattern where larger harvest invariably led to lower prices. It may be that demand has been stimulated since the increased production of maize has helped reduce poverty. There remains the possibility that harvests have been over-estimated.
5. Subsidies on triple super-phosphate and muriate of potash are even larger, since these also sell at just over USD 3 a bag, but are more expensive on world markets than urea.

References

Ariga, J. and T.S. Jayne (2009), "Unlocking the market: Fertilizer and maize in Kenya", Chapter 14 in D. Spielman and R. Pandya-Lorch (eds.), *Millions Fed: Proven Successes in Agricultural Development*, International Food Policy Research Institute, Washington, DC.

Chand, R. and P. Kumar (2004), "Determinants of capital formation and agriculture growth. some new explorations", *Economic and Political Weekly*, Vol.39, pp. 5611-5616.

Crawford, E.W., T.S. Jayne and V.A. Kelly (2005), "Alternative Approaches for Promoting Fertilizer Use in Africa, with Particular Reference to the Role of Fertilizer Subsidies", Paper, Department of Agricultural Economics, Michigan State University, East Lansing, MI.

Dorward, A. and E. Chirwa (2011), "The Malawi agricultural input subsidy programme: 2005-6 to 2008-9", *International Journal of Agricultural Sustainability* (IJAS), January 2010, Vol. 9.

Dorward, A., S. Fan, J. Kydd, H. Lofgren, J. Morrison, C. Poulton, N. Rao, L. Smith, H. Tchale, S. Thorat, I. Urey and P. Wobst (2004), "Institutions and policies for pro-poor agricultural growth", *Development Policy Review*, Vol. 22, pp. 611-622.

Dorward. A. (2009), "Rethinking Agricultural Input Subsidy Programmes in a Changing World", Draft for FAO, February.

Duflo, E., M. Kremer and J. Robinson (2008), "How high are rates of return to fertilizer? Evidence from field experiments in Kenya", *American Economic Review Papers (Papers and Proceedings Issue)* Vol. 98, pp. 482-488.

Ekanayake, H.K.J. (2006), "The impact of fertilizer subsidy on paddy cultivation in Sri Lanka", *Staff Studies*, Vol. 36 (1 and 2), pp. 73-101.

FAC (2008), "The Malawi fertiliser subsidy programme: Politics and pragmatism", *Briefing Paper* by Blessings Chinsinga, February 2008, Future Agricultures Consortium, Brighton, United Kingdom.

FAC (2009), "Fertiliser subsidies: Lessons from Malawi for Kenya", *Briefing Paper*, February 2009, Future Agricultures Consortium, Brighton United Kingdom.

Heisey, P.W. and G.W. Norton (2007), "Fertilizers and other farm chemicals", *Handbook of Agricultural Economics*, Elsevier.

Kikuchi, Masao and P.B. Aluwihare (1990), "Fertilizer response functions of rice in Sri Lanka: Estimation and some application", International Irrigation Management Institute, Colombo, Sri Lanka.

Mullen, Kathleen, David Orden and Ashok Gulati (2005), "Agricultural policies in India. Producer Support Estimates 1985-2002", *MTID Discussion Paper* 82, International Food Policy Research Institute, Washington, DC.

OECD (2001), *Market Effects of Crop Support Measures*, OECD, Paris.

Smith, L.E.D. and I. Urey (2002), "Agricultural Growth and Poverty Reduction: A Review of Lessons from the Post-Independence and Green Revolution Experience in India", *Report* written as part of a research project on Institutions and Economic Policies for Pro-poor Agricultural Growth, funded by the Department for International Development of the United Kingdom (ESCOR Project R7989), Wye, Kent: Department of Agricultural Sciences, Imperial College at Wye.

Tibbotuwawa, M. (2010), "Better targeting of transfers: The fertilizer subsidy", Talking Economics blog, Monday, 16 August 2010: at *http://ipslk.blogspot.com/2010/08/better-targeting-of-transfers.html.*

World Bank (2003), "Sri Lanka promoting agricultural and rural non-farm sector growth", *Report* No. 25387-CE, World Bank, Washington, DC.

World Bank (2007), *World Development Report 2008; Agriculture for Development*, World Bank, Washington, DC.

ORGANISATION FOR ECONOMIC CO-OPERATION AND DEVELOPMENT

The OECD is a unique forum where governments work together to address the economic, social and environmental challenges of globalisation. The OECD is also at the forefront of efforts to understand and to help governments respond to new developments and concerns, such as corporate governance, the information economy and the challenges of an ageing population. The Organisation provides a setting where governments can compare policy experiences, seek answers to common problems, identify good practice and work to co-ordinate domestic and international policies.

The OECD member countries are: Australia, Austria, Belgium, Canada, Chile, the Czech Republic, Denmark, Estonia, Finland, France, Germany, Greece, Hungary, Iceland, Ireland, Israel, Italy, Japan, Korea, Luxembourg, Mexico, the Netherlands, New Zealand, Norway, Poland, Portugal, the Slovak Republic, Slovenia, Spain, Sweden, Switzerland, Turkey, the United Kingdom and the United States. The European Commission takes part in the work of the OECD.

OECD Publishing disseminates widely the results of the Organisation's statistics gathering and research on economic, social and environmental issues, as well as the conventions, guidelines and standards agreed by its members.

OECD PUBLISHING, 2, rue André-Pascal, 75775 PARIS CEDEX 16
(51 2011 08 1 P) ISBN 978-92-64-16863-3 – No. 59749 2012-02